The MIDI Manual

The MIDI Manual

David Miles Huber

SAMS

A Division of Macmillan Computer Publishing

11711 North College, Carmel, Indiana 46032 USA

FIRST EDITION
FIRST PRINTING—1990

International Standard Book Number: 0-672-22757-6
Library of Congress Catalog Card Number: 90-63630

Acquisitions and Development Editor: *James Rounds*
Manuscript Editor: *Judy Brunetti*
Technical Reviewers: *Rick Davies and Gary Lindsay*
Cover Design: *Glen Santner*
Illustrator: *T. R. Emrick*
Production Coordinator: *Sally Copenhaver*
Production: *Jeff Baker, Claudia Bell, Martin Coleman, Tami Hughes, Chuck Hutchinson, Betty Kish, Bob LaRoche, Sarah Leatherman, Howard Peirce, Cindy L. Phipps, Joe Ramon, Dennis Sheehan, Suzanne Tully, Lisa Wilson*
Indexer: *Sharon Hilgenberg*

Printed in the United States of America

This book is dedicated to Bruce Hamerslough, Steve Royea, and (last but certainly not least) Walt Wagner for letting me bend their ears during the course of this book's production. It is also dedicated to all of the women who are pioneering within this techno-artistic field.

Contents

Foreword

Throughout history the tools used by artists, such as painters, sculptors, writers, and musicians have undergone continual change. Some changes have been more significant than others. However, all of them have advanced the artistic process in one way or the other.

As musicians living and working in the past decade we have had the unique opportunity to experience one of the most radical changes in the tools we use to occur in the history of music making. Ranking right up there with the development of the pipe organ and sound recording, the integrating of computers with musical instruments has, once again, changed the way music will be made forever. The impetus behind this surge was the development in 1983 of the MIDI interface. With this new technology musicians could now interconnect virtually all the elements of their studio and use the power of the computer to manage all the power that was suddenly at their fingertips.

As revolutionary as the devlopment of MIDI was, it is no longer a new and novel technology. MIDI has gone the way of all significant changes. It has become commonplace; arguably the greatest achievement for any new development. However, the more entrenched something becomes, the more need there is to understand and master it—and this is not always an easy task. Virtually every musician has been familiarized with the piano keyboard from the point they began studying music. It is a musical standard that has been around for generations. Unfortunately, we have not had the luxury of growing up with MIDI (although for many 20-year-olds this may not be the case) which necessitates continuing education for forward-minded musicians to broaden their understanding beyond their particular instrument of choice. The better you understand your tools the more energy you can put into making music.

Does this mean that to become a musician, you have to first study computer science? Absolutely not. The best technology is one that you don't have to think about, and MIDI can function in that capacity. However, by learning more about what MIDI can do and how it does it, you often expand your capabilities—automatically adjusting the volume of your synth via MIDI will allow not having to think about setting the synth up each time you play the song—and give you more time to create music.

So, read this book and learn about all the realities and possibilities that MIDI has provided the modern musician. And then, get on with making music!

LACHLAN WESTFALL

Preface

Both MIDI (Musical Instrument Digital Interface) and electronic musical instruments have had a major impact upon the techniques and production styles of modern music, live stage production, video, and film. With this in mind, and as the adoption of the MIDI 1.0 Specification moves into its second decade, there is a strong need for what I call "second generation MIDI books." These are books which cover topics that relate to the most recent advances, hardware, and technology within MIDI and electronic music production. *The MIDI Manual* responds to this need by being the first of these second generation books to introduce the reader to the latest industry advances, while thoroughly covering the established basics of MIDI technology.

Topics which are covered within this book include:

- A brief history of MIDI and its applications to the music, video, film, and live performance industries.
- The digital word and a detailed introduction to the structure of the MIDI message.
- System interconnections that are required within a MIDI system, hardware for managing MIDI data, as well as various personal computers and their applications within the MIDI setup.
- The various electronic musical instruments, including controllers, keyboard instruments, percussion, MIDI guitar systems, and MIDI woodwind systems.
- Sequencing with a detailed overview of the various sequencing system types.
- MIDI patch editor and librarian programs.
- Music-printing programs.
- MIDI-based signal processors.
- Synchronization, including synchronization between audio transports, non-MIDI sync, and the various MIDI sync standards and hardware systems.

- Digital audio technology as it applies to MIDI production, MIDI-based mixing, and automation.
- The MIDI 1.0 and MIDI time-code (MTC) specifications.

The MIDI Manual was written as a reference or text book for those who are currently involved within modern music production; from the curious newcomer to the established professional artists. It is also intended for those people within the video, film, and live-stage media, who would like to gain insights into how MIDI effects each of these industries.

The present and potential power of MIDI has had me psyched for quite a while, and I hope that this book will both excite and inform you about how MIDI can best serve your production needs. I'll close with an old (and slightly off-the-wall) Confucianist proverb, "Music and its emotion cannot exist in 'chips,' but can only be found in the expression and creativity of the human heart." Dive off the deep end and have fun!

Happy trails,

DAVID MILES HUBER

Acknowledgments

I would like to thank the following individuals and companies who have assisted in the preparation of this book by providing photographs and technical information: Becky David and Chuck Tompson of J.L. Cooper Electronics; Leah Holsten of Lexicon Inc; Jeff Klinedinst of Turtle Beach Softworks; John Mavraides of Mark of the Unicorn, Inc.; Harry E. Adams of Adams-Smith; Caroline Meyers of Intelligent Music; Carmine Bonanno of Voyetra Technologies; James Martin of Akai Professional/IMC; Jeffrey J. Mercer of Tsunami Software; Doria A. Marmaluk of Ensoniq Corp.; Dennis Briefel of Music Industries Corp.; Kate Prohaska-Sullivan of Commodore Business Machines, Inc.; Suzette Mahr of Words & Deeds, Inc.; Mark W. Grasso of Magnetic Music; Julie Payne-McCullum of Coda Music Software; Linda Petrauskas of E-mu, Inc.; Jim Fiore and Rob Lewis of Dissidents; Lori W. Bradford of Lake Butler Sound Company; Melinda Turscanyi of Musicode; Paul De Benedictis and Kord Taylor of Opcode Systems, Inc.; Jim Giordano of Studiomaster; Larry Winter of Rane Corporation; Dan Hergert of Kawai America Corporation; Nigel J. Redmond of EarLevel Engineering; David Rowe of Music Quest, Inc.; Larry DeMarco of Korg USA, Inc.; Charles Henderson of Current Music Technology; F. Stuckey McIntosh of Gambatte Digital Wireless, Inc.; Marci Galea of Electronic Arts; Paula Lintz and Marvin Ceasar of Aphex Systems; Don Bird of 360 Systems; Steve Salani of Forte Music; Gerry Tschetter and Phil Moon of Yamaha Corporation of America; Russ Kozerski of Cool Shoes Software; Jeff Sorna of Rocktron Corporation; William C. Mohrhoff of Tascam; Myra B. Arstein of Optical Media International; Maria S. Katoski of Kat, Inc.; Tony Lauria of Cannon Research Corporation; Scott Berdell of Quest Marketing; Glenn J. Hayworth of Sound Quest, Inc.; Rich Rozmarniewicz and Laura Barnes of Eltekron Technologies, Inc.; John Parker of Dynacord Electronics Inc.; Scott Berdell of Quest Marketing; Chris Ryan of ddrum; Gene Barkin of Digital Music Corp.; Michael Amadio of Cordata Technologies, Inc.; Janet Susich of Apple Computer, Inc.; Al Hospers of Dr. T's Music Software; Chez C. Bridges of Hybrid Arts, Inc.; Anastasia Lanier of Passport Designs, Inc.

I would also like to express my special thanks to my friend and developmental editor, Jim Rounds, my manuscript editor, Judy Brunetti, and to Rick Davies for his excellent job as technical editor.

Trademark Acknowledgments

Chapter 1

Introduction

Simply stated, *Musical Instrument Digital Interface (MIDI)* is a digital communications language and compatible hardware specification that allows multiple electronic instruments, performance controllers, computers, and other related devices to communicate with each other within a connected network (Fig. 1-1). It is used to translate performance- or control-related events (such as the actions of playing a keyboard, selecting a patch number, varying a modulation wheel, etc.) into equivalent digital messages and to transmit these messages to other MIDI devices where they can be used to control the sound generation and parameters of such devices within a performance setting. In addition, MIDI data can be recorded into a digital device (known as a *sequencer*), which can be used to record, edit, and output MIDI performance data.

In artistic terms, this digital language is a new and emerging medium that enables artists to express themselves with a degree of expression and control that was, until recently, not possible on an individual level. Through the transmission of this digital performance language within an electronic music system, a musician can create and develop a song or composition in a practical, flexible and affordable production environment. In addition to composing and performing a song, the musician is also able to act as a techno-conductor, having complete control over the wide range of individual sounds that are created by various instruments, their *timbre* (sound and tonal quality), and *blend* (level, panning, and other real-time controls). MIDI may also be used to vary the performance/control parameters of both electronic instruments and effects devices during a performance.

The term *interface* refers to the actual data communications link and hardware that occurs in a connected MIDI network. MIDI makes it possible for all electronic instruments and devices to be addressed upon this network through the transmission of real-time performance and control-related messages. Furthermore, a number of instruments or individual sound generators within an

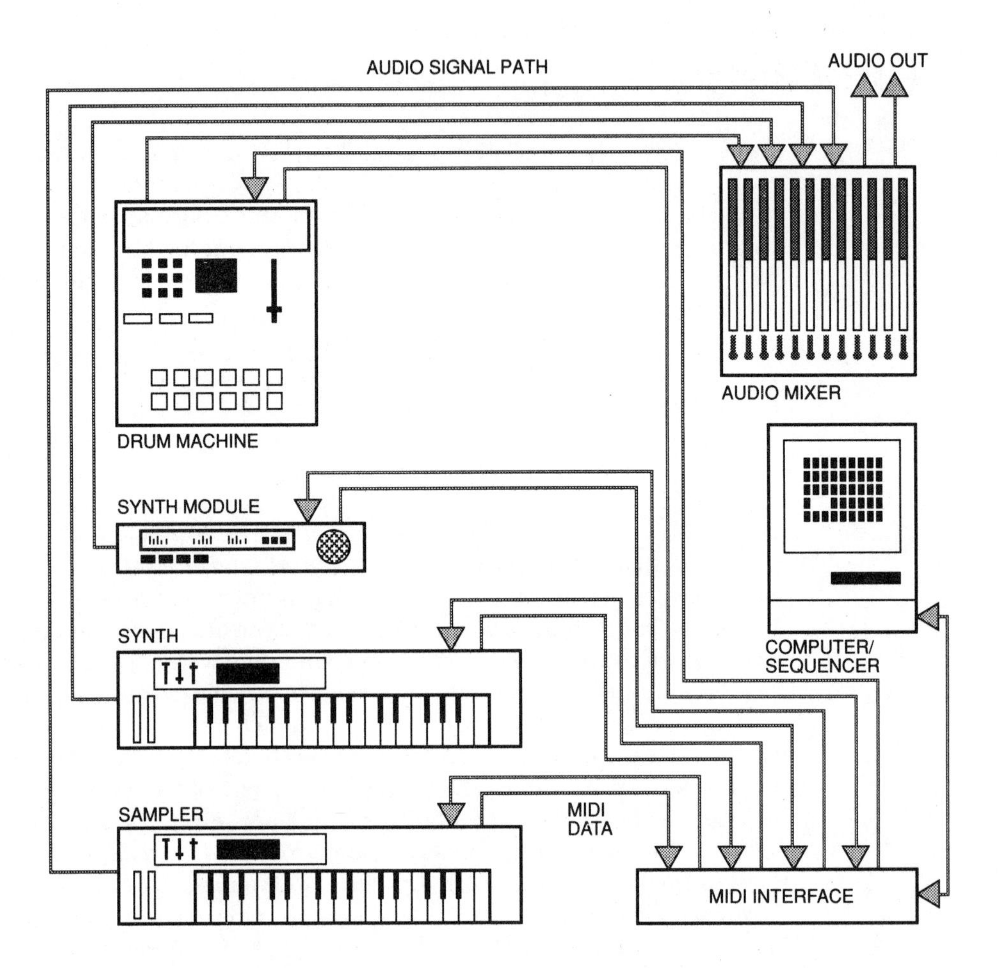

Fig. 1-1. Example of a typical MIDI system. (*Courtesy of J. Rona Music*)

instrument can be independently addressed upon a single MIDI line. This is possible as each MIDI data line is capable of transmitting performance and control messages over 16 discrete channels. This multichannel arrangement allows the performer to record, control, and reproduce MIDI data in a working environment that resembles the multitrack recording process which is familiar to most musicians. Once mastered, however, MIDI offers production challenges and possibilities that are beyond the capabilities of the traditional multitrack recording studio.

A Brief History

In the early days of electronic music, keyboard synthesizers were commonly *monophonic devices* (capable of sounding only one note at a time) and often generated a thin sound quality. These limiting factors caused early manufacturers to look for ways to link more than one instrument together to create a thicker, richer sound texture. This could be accomplished by establishing an instrument link that would allow one synthesizer (acting as a master controller) to be used to directly control the performance parameters of one or more synthesizers (known as *slave sound modules*). As a result, a basic control signal (known as *control voltage* or *CV*) was developed (Fig. 1-2).

This system was based upon the fact that when most early keyboards were played, they generated a DC voltage that was used to directly control voltage-controlled oscillators (which affected the pitch of a sounding note) and voltage-controlled amplifiers (which affected the note's volume and on/off nature). Since many keyboards of the day generated a DC signal that ascended at a rate of 1 volt per octave (breaking each musical octave into 1/12-volt intervals), it was possible to use this standard control voltage as a master-reference signal for transmitting pitch information to additional synthesizers. In addition to a control voltage, this standard required that a keyboard transmit a *gate signal*. This second signal was used to synchronize the beginning and duration times which are to be sounded by each note. With the appearance of more advanced *polyphonic synthesizers* (capable of generating multiple sounds at a time) and early digital devices on the market, it was clear that this standard would no longer be the answer to system-wide control, and new standards began to appear on the scene (creating incompatible control standards). With the arrival of early drum machines and sequencing devices, standardization became even more of a dilemma.

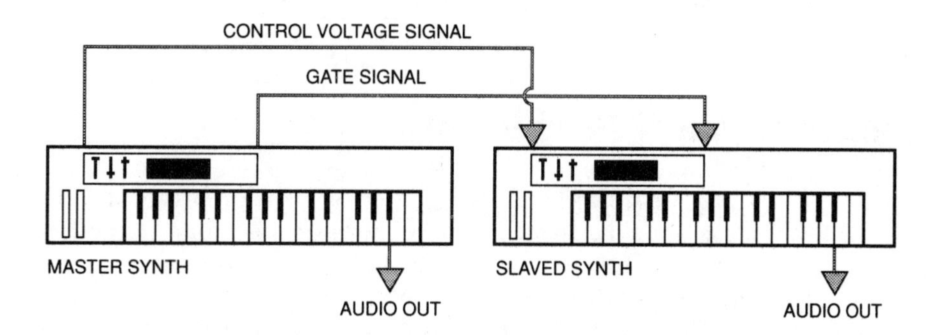

Fig. 1-2. Example of an instrument link using a control voltage and gate signal.

Synchronization between such early devices was often problematic, as manufacturers would often standardize upon different sync-pulse clock rates. These timing incompatibilities would result in a loss of sync over a very short time (when locking between devices from different manufacturers). This would render synchronization nearly impossible without additional sync-rate converters or other types of modifications.

As a result of these incompatibilities, Dave Smith and Chet Wood (then of *Sequential Circuits*, an early manufacturer of electronic instruments) began creating a digital electronic instrument protocol, which was named the *Universal Synthesizer Interface* (*USI*). As a result of this protocol, it was possible for equipment from different manufacturers to directly communicate with one another. For example, the synthesizer from one company would work with another company's sequencer. In the fall of 1981, USI was proposed to the Audio Engineering Society. Over the course of the following two years, a panel (which included representatives from the major electronic instrument manufacturers) modified this standard and adopted it under the name of Musical Instrument Digital Interface (MIDI Specification 1.0).

The acceptance of MIDI was largely due to the strong need for a standardized protocol and advances in technology that allowed complex circuit chips and hardware design to be cost-effectively manufactured. It was also due, in part, to the introduction of Yamaha's DX-7 synthesizer in the winter of 1983, after which time keyboard sales began to grow at an astonishing rate.

With the adoption of this industry-wide standard, any device that incorporates a series of MIDI ports into its design is capable of transmitting and/or responding to the digital performance and control-related data that conforms to the MIDI specification. As the vast majority of electronic instruments and devices implement this communications language, the electronic musician is assured that the basic functions of any new device will integrate into an existing MIDI system.

Electronic Music Production

Today MIDI systems are being used by many professional and nonprofessional musicians alike. In turn, these systems are relied upon to perform an expanding number of production tasks, including music production, audio-for-video and film post-production, stage production, etc.

This industry acceptance can, in large part, be attributed to the cost-effectiveness, power, and general speed of MIDI production. Currently, a large variety of affordable MIDI instruments and devices are available on the market. Once purchased, there is often less need (if any at all) to hire outside musicians for a production project. In addition, MIDI's multichannel production environment enables a musician to compose, edit, and arrange a piece with a high degree of flexibility, without the need for recording and overdubbing these sounds onto multitrack tape.

This affordability potential for future expansion and increased control capabilities over an integrated production system has spawned the growth of an industry, which is also very personal in nature. For the first time in music history, it is possible for an individual to realize a full-scale sound production in a cost- and time-effective manner. As MIDI is a real-time performance medium, it is also possible to fully audition and edit this production at every stage of its development—all within the comfort of your own home or personal production studio.

MIDI systems may also be designed in any number of personal configurations. They can be designed to best suit a production task (the production of a video sound track), to best suit the main instrument and playing style of an artist (percussion, guitar, keyboards), or to best suit a musician's personal working habits (choice of equipment and/or design layout). Each of these advantages are a definite tribute to the power and flexibility that is inherent within the capabilities that MIDI brings to modern music production.

MIDI in the Home

There is currently a wide range of electronic musical instruments, effects devices, computer systems, and other MIDI related devices that are available within the electronic music market. This range of diversity allows the electronic musician to select various devices which best suit his or her particular musical taste and production style. With the advent of the *large-scale integrated circuit chip* (*LSI*), which enables complex circuitry to be quickly and easily mass-produced, many of the devices that make up an electronic music system are affordable by almost every musician or composer, whether they are a working professional, aspiring artist, or hobbyist. As most of these devices are digital or digitally controlled in nature, it can often be a simple and cost-effective matter to implement them with MIDI capabilities.

Due to the personal nature of MIDI, a production system can show up in a number of environments and can be configured to perform a wide range of applications. Basic and even many not-so-basic MIDI systems are currently being installed within the homes of a growing number of electronic musicians. These can range from systems which take up a corner of an artist's bedroom to larger systems that have been installed within a dedicated MIDI production studio.

Home MIDI production systems can be designed to functionally serve a wide range of applications, and they can be used by hobbyists and aspiring or professional musicians alike. Such systems also have the important advantage of allowing the artist to produce his or her music in a comfortable environment (whenever the creative mood hits). Such production luxuries, which would have cost an artist a fortune only a decade ago, are now within the reach of nearly every musician.

(A) J. Rona Music. (Courtesy of Jeff Rona and Home & Studio Recording Magazine)

(B) Walt Wagner Music Productions. (Courtesy of Walt Wagner)

Fig. 1-3. Examples of the MIDI production studio.

MIDI production systems often appear in a wide range of shapes and sizes and can be designed to match a wide range of production and budget needs. For example, a keyboard instrument (commonly known as a *MIDI workstation*), will often integrate a keyboard, polyphonic synthesizer, percussion sounds, and a built-in sequencer into a single package. This is essentially an all-in-one system that allows a musician to cost-effectively break into MIDI style production. Additional MIDI instruments can easily be added to this workstation by simply plugging its MIDI out port to the MIDI in port of the new device (or devices), plug the extra audio channels into your mixer, and you're in business.

Other MIDI systems, made up of discrete MIDI instruments and devices, are often carefully selected by the artist to generate a specific sound or to serve a specific function within the system. Although this type of MIDI system is often not very portable, it is commonly more powerful, as each component is designed to perform a wider range of task-specific functions. Such a system might include one or more keyboard synthesizers, synth modules, samplers, drum machines, a computer (with sequencer and other forms of MIDI software), effects devices, and audio mixing capabilities.

MIDI in the Studio

MIDI has also dramatically changed the sound, technology, and production habits within the professional recording studio. Before the onset of MIDI, the recording studio was one of the only production environments that would allow an artist or composer to combine various instruments, styles, and timbres into a final recorded product. The process of recording a group in a live setting, often was (and is) an expensive and time consuming undertaking. This is due to the high cost of hiring session musicians and the high hourly rates charged for the use of a professional music studio. In addition, a number of hours are usually spent capturing the "ideal performance" on tape.

With the advent of MIDI, a large part of modern music production can be performed and, to a large extent, produced before going into the recording studio. This capability has reduced the number of hours required for laying tracks on a multitrack tape to a cost-effective minimum. For example, it is now commonplace for groups to preproduce an entire album project by using their own personal MIDI system. Once completed (or nearly completed), the group can bring the MIDI system into the studio and lay the isolated audio channels down to a multitrack recorder. Once all of the MIDI tracks have been transferred to tape, the process of recording vocals, live sources, or flying in vocal samples to tape can begin. Afterwards, the completed master tape can be mixed down into a final musical product. Alternatively, the MIDI tracks may not be recorded onto multitrack tape at all, but may be synchronized to tape, allowing the sequencer to act as an extension of the recorded tracks.

MIDI within Audio-For-Video and Film

Electronic music has also become a major tool within the scoring and audio postproduction phases for the professional video and film media. Standard MIDI production practices have enabled music composers and effects artists for the visual media to become increasingly productive. As a result, MIDI has come to be recognized as a powerful, cost-effective tool in the production of TV commercials, industrial videos, TV, and full-feature motion picture sound tracks.

Although many electronic music composers for the visual media work independently, many have teamed up with recording studios, video production houses, and audio-visual firms to combine forces in creating a full in-house visual and music production team.

MIDI in Live Performance

Electronic music production and MIDI are also firmly established within the stage setting of live performance. In addition to the sound quality produced by many modern electronic keyboard systems, MIDI's onstage popularity is primarily due to two factors: preproduction programming and the ability to easily control a number of devices from a central location.

The ability to sequence a number of rhythm and background parts in advance, chain them together into a single, controllable sequence (using a jukebox-type sequencing program), and play back this sequence onstage has become a strong live-performance tool for many musicians. This technique is currently in use by solo artists who have become one-man-bands by adding background sequences (often consisting of a drum machine and a host of other possible instruments) to their live vocals and playing. Larger techno-pop groups commonly make use of extensive onstage sequencing for driving a wide range of instruments and effects in addition to their live playing.

Many live performers make use of the real-time control capabilities of MIDI in an onstage environment. This includes the ability of a number of slaved devices to be played from a single keyboard, percussion pad, or other type of controller. MIDI also allows different musical and percussion parts to be played from a single controller. In addition to communicating performance-related data, MIDI is capable of controlling a number of device parameters in real time. For example, a central MIDI controller (such as a keyboard controller or MIDI footswitch controller) can be used to change the sound patches and performance parameters (such as volume, panning, etc.) of various MIDI instruments or effects devices.

MIDI is also indirectly involved in the creation of many live performances. Through the use of a music notation program and production system, MIDI can often help a composer input, edit, and audition a musical score. Once finalized, a sequenced computer score or series of lead sheets can be printed out as hard copy for distribution to the live players onstage or in the studio.

Summary

From these applications, it can be seen that Musical Instrument Digital Interface (MIDI) has become widely accepted by consumers and professionals in many facets of the music industry. It has also brought about sweeping changes in the way that many styles of music are produced, making it possible for the individual artist to create his or her music in an affordable and practical manner. MIDI has accomplished this with an amazing degree of success by effectively integrating the electronic music production system. Through the use of MIDI, it is possible for any controlling source (such as a keyboard, drum pads, etc) to directly communicate with another instrument within the system. Using such an arrangement, it is a simple matter to slave more than one instrument to a single controller (thereby creating a richer, more complex sound texture, and reducing the need for additional redundant keyboards).

Through the digital communications language of MIDI it is also possible for electronic musical instruments and devices to directly communicate with a personal computer. This allows the artist to utilize the processing power of the personal computer for performing a wide range of recording, editing, librarian, and control tasks via MIDI.

Chapter 2

Musical Instrument Digital Interface

The musical instrument digital interface is a digital communications protocol. That is to say, it is a standardized control language and hardware specification that allows multiple electronic musical instruments and devices to communicate real-time and nonreal-time performance and control data.

What Is MIDI?

MIDI is a specified data format that must be strictly adhered to by those who design and manufacture MIDI equipped instruments and devices. In this way, task-related performance and automation messages can be communicated between devices with relative transparency, speed, and ease. Thus, when performing a task (such as controlling multiple instruments from a keyboard controller or transmitting a patch bank from a patch librarian to a synthesizer), the user need only consider the control parameters of the involved devices and not those of the transmission medium itself. This could be likened to our English language, through which ideas can be transmitted from one person to an audience. For example, as English speaking people we are able to concentrate wholly upon the content of a lecture without having to think about the language medium itself. Similarly, performance and control data (between task-compatible devices) are able to communicate through the same chosen communications medium of MIDI.

The Digital Word

One of the best ways to gain insight into the MIDI specification is to expand on the previous example of comparing MIDI to a spoken language.

As humans, we have adapted our communication skills to best suit our physical selves. Ever since the first grunt, we have found it easiest to communicate through the use of our vocal chords, and we've been doing it ever since that time, when we found out that other people or animals could hear our grunts and respond to them (hopefully positively). Thus, over time, language was developed which assigned a standardized meaning to a series of sounds (words). Eventually these words were grouped together to convey a more complex means of communication. In order to record our English language, a standard method of notation was developed that assigned 26 symbols to these sounds (letters of the alphabet), which, when written as a group, would communicate an equivalent spoken word (Fig. 2-1). By stringing these written words into complete sentences, more complex communication could be recorded for later review. For instance, the letters T, I, & E don't mean much when used individually. However, when they're grouped into a word, they refer to a piece of cloth that is worn around the neck as a social convention. When placed within a sentence, words are given a greater clarity of meaning (e.g., "There's a guy with a bow tie.").

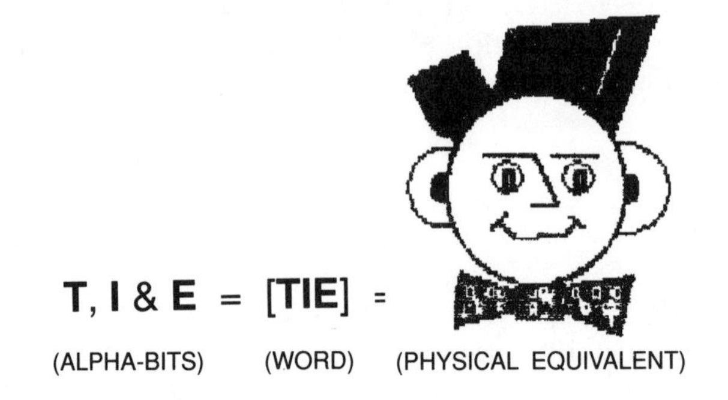

Fig. 2-1. Meaning is given to the alphabets T, I, & E, when they are grouped into a word and placed within a sentence.

On the other hand, microprocessors and computers are digitally based communications devices that obviously lack vocal chords and ears (although even that's changing). However, they do have the unique advantage of being able to process numbers at a very high rate. Unlike our base-10 system of counting, computers are limited to communicating within the binary system of 0's and 1's (or on and off). Like humans, most computers are capable of grouping these *bi*-nary dig-*its* (known as ***bits***) into larger numeric values. These bits can then be grouped together to form a digital word which can be used to represent and communicate specific information and instructions. Just as a human can communicate a simple sentence, a computer is able to generate and respond to a series of related digital words that are understood by a digital system (Fig. 2-2). The rate at which this data can be communicated is measured in *baud* or the number of 8-bit words that are transmitted or received per second.

(1001 0100) (0100 0000) (0101 1001)

(STATUS BYTE) (DATA BYTE #1) (DATA BYTE #2)

Fig. 2-2. General example of a digitally generated MIDI message.

The MIDI Message

Musical performance data is communicated digitally throughout a production system as a string of MIDI messages which is transmitted through a single MIDI line at a speed of 31.25 kbaud. Data within a single MIDI line can only travel in one direction, from a single source to a destination (Fig. 2-3A). In order to make two-way communication possible, a second MIDI data line must be connected from the slave source back to our original master device (Fig. 2-3B).

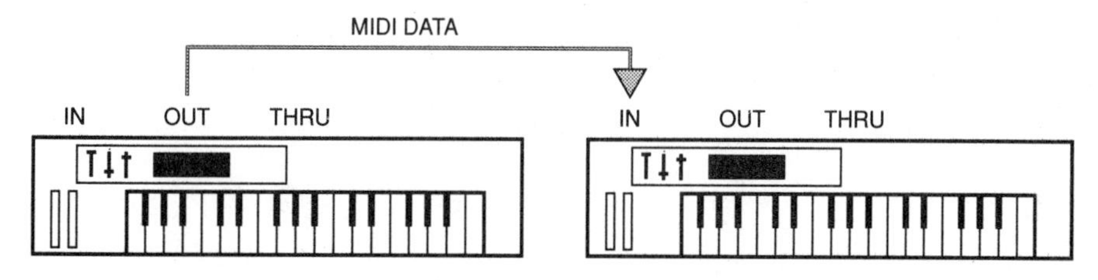

(A) Data transmission from a single source to a destination.

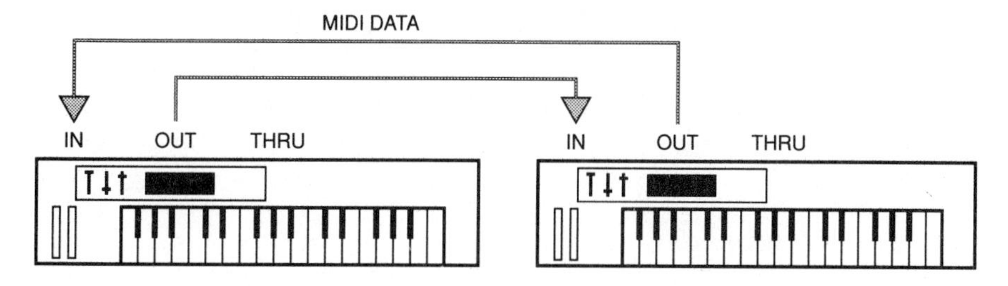

(B) Two-way communication using two MIDI cables.

Fig. 2-3. MIDI data may only travel in one direction through a single MIDI cable.

MIDI messages are made up of a group of 8-bit words (known as *bytes*), which are transmitted in a serial fasion to convey a series of instructions to one or all MIDI devices within a system.

There are only two types of bytes that are defined within the MIDI specification: the *status byte* and the *data byte*. Status bytes are used within the MIDI message as an identifier for instructing the receiving device to which particular MIDI function and channel are being addressed. The data byte is used to encode the actual numeric value which is to be attached to the accompanying status byte. Although a byte is made up of 8 bits, the *most significant bit* (*MSB*) (the left-most binary bit within a digital word) is used solely to identify the byte type. The MSB of a status byte is always 1, while the MSB of a data byte is always 0 (Fig. 2-4). For example, a 3-byte MIDI note-on message (which is used to signal the beginning of a MIDI note) in binary form might read:

	Status Byte	Data Byte 1	Data Byte 2
Description	Status/Channel #	Note #	Attack Velocity
Binary Data	(1001 0100)	(0100 0000)	(0101 1001)
Numeric Value	(Note On/Ch. #4)	(64)	(89)

Thus, a 3-byte note-on message of (10010100) (01000000) (01011001) would transmit instructions that would read:

Transmitting a note-on message over MIDI channel #4, for note #64, with an attack velocity (volume level of a note) of 89.

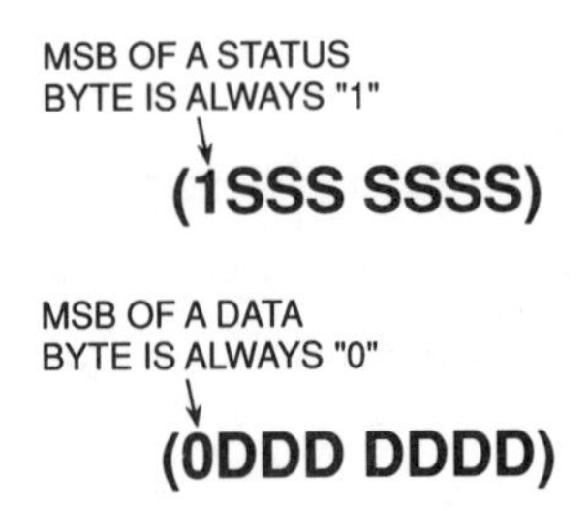

Fig. 2-4. The most significant bit of a MIDI data byte is used to identify between a status byte 1 and a data byte 0.

MIDI Channels

Just as it is possible for a public speaker to single out and communicate a message to one individual within a crowd, MIDI messages may be directed to a specific device or range of devices. This is done by including a nibble (four bits) within the status/channel number byte (Fig. 2-5) which instructs all receiving devices to which MIDI channel a specific message is being transmitted upon. As the channel nibble is 4-bits wide, up to 16 channels can be transmitted through a single MIDI cable.

FINAL 4-BITS "NIBBLE"
OF A STATUS BYTE IS
USED TO ENCODE THE
MIDI CHANNEL NUMBER

(1SSS CCCC)

Fig. 2-5. The least significant nibble of the status/channel number byte is used to encode the channel number.

Whenever a MIDI device is instructed to respond to a specific channel number, it will ignore channel messages that are transmitted upon any other channel. Likewise, any device that is set to respond to a specific MIDI channel will only respond to messages that are transmitted upon that channel (within the capabilities of the device).

For example, let's assume that we have two synthesizers and a *MIDI sequencer* (a device that is capable of recording, editing, and outputting MIDI data) with which to create a short song. We might start off by playing a melody line into our sequencer on Synthesizer A, which is set to transmit and respond to data upon MIDI Channel #3 (Fig. 2-6). Having done this, we can then decide to play background chords upon Synthesizer B, which we shall set to MIDI Channel #4. Even though the system is connected by one MIDI line, it would be a simple matter for our sequencer to output the previously recorded MIDI data on Channel #3 (which will still be played by Synth A), while our Synth B simultaneously responds to our live playing of the keyboard. Upon playing back the finished MIDI sequence, both of our instruments will respond only to their assigned MIDI channels and will reproduce their individual sounds as they were recorded.

MIDI Modes

Electronic instruments often vary in the number of sounds that can be produced at one time using their internal sound-generating circuitry. They may also vary in the number of individual characteristic sounds that can be simultaneously produced by an instrument. For example, certain instruments are only capable of producing one note at a time, while others (known as *polyphonic instruments*) are capable of generating numerous notes at one time. The latter allows the artist to play chords and more than one musical line upon a single instrument. In addition, it is possible for various synthesizer types to produce only one characteristic sound patch at a time (e.g., electric piano, synth bass, etc). The word *patch* is a direct reference to the need for *patch chords* when using earlier analog synthesizers to connect one sound generator or processor with another. On the other hand, a single instrument may also be *multitimbral* in nature, meaning that it is capable of generating more than one sound patch at a time.

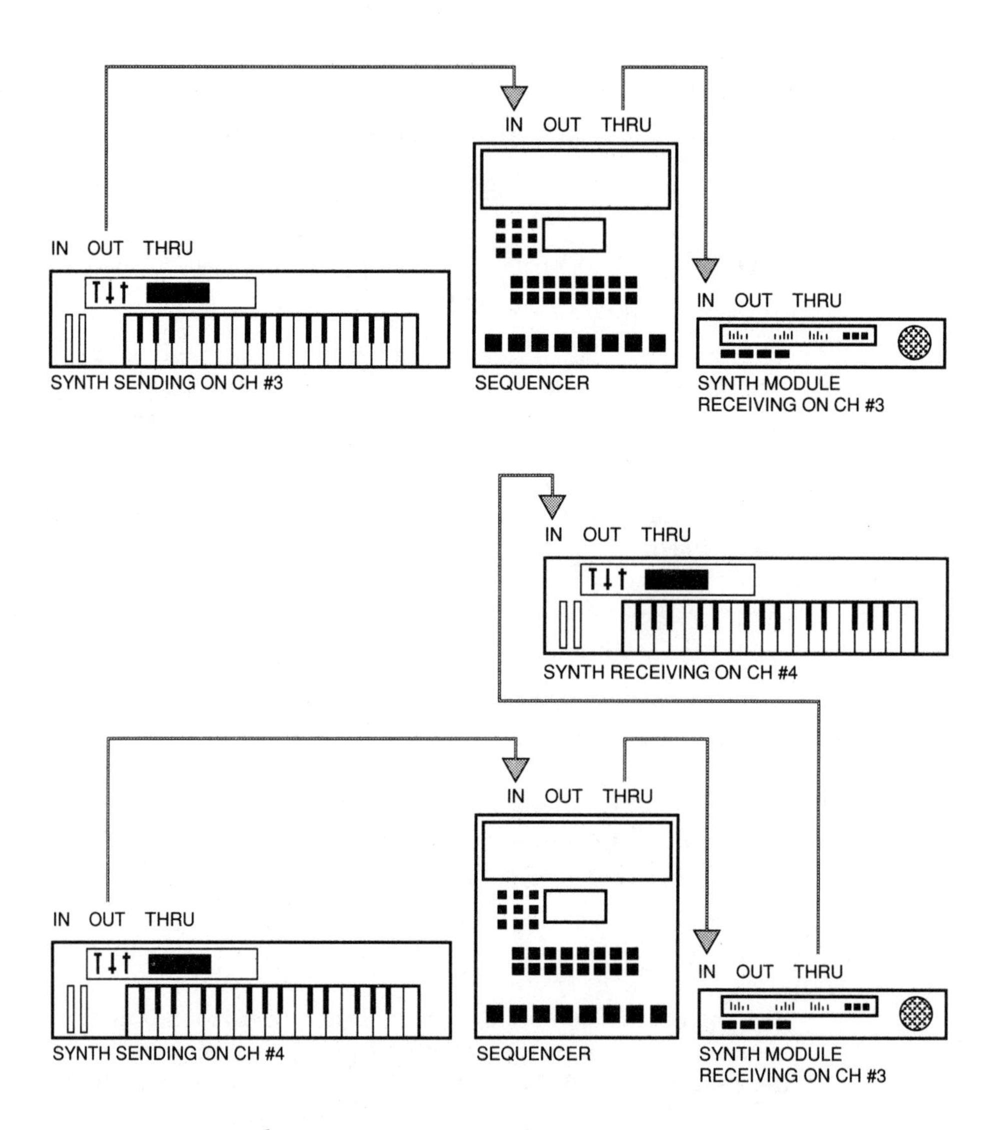

Fig. 2-6. System showing a set of MIDI channel assignments.

As a result of these differences between devices, a defined set of guidelines (known as *MIDI reception modes*) has been specified which allows a MIDI instrument to transmit or respond to MIDI channel messages in several ways. For example, an instrument might be programmed to respond to all 16 MIDI channels at one time, while another might be polyphonic in nature and programmed to respond to only one MIDI channel. It is also common for a single instrument to be both polyphonic and multitimbral (allowing a number of generated sound patches to individually respond to their own assigned MIDI channel). A listing of these modes are:

Mode 1 — Omni on/poly
Mode 2 — Omni on/mono
Mode 3 — Omni off/poly
Mode 4 — Omni off/mono

Omni on/off refers to how a MIDI instrument will respond to the 16 MIDI channels. When Omni is turned *on*, the MIDI device will respond to all channel messages that are transmitted over all MIDI channels. Whenever Omni is turned *off*, the device will only respond to a single MIDI channel or set of assigned channels. *Poly/mono* refers to the sounding of individual notes by a MIDI instrument. In the *poly mode*, an instrument is capable of responding polyphonically to each MIDI channel and is able to produce more than one note at a time. In the *mono mode*, an instrument is capable of responding monophonically to each MIDI channel and is capable of producing only one note at a time).

- *Mode 1 — Omni on/poly*: An instrument will be able to respond polyphonically to performance data that is received upon any MIDI channel (Fig. 2-7).

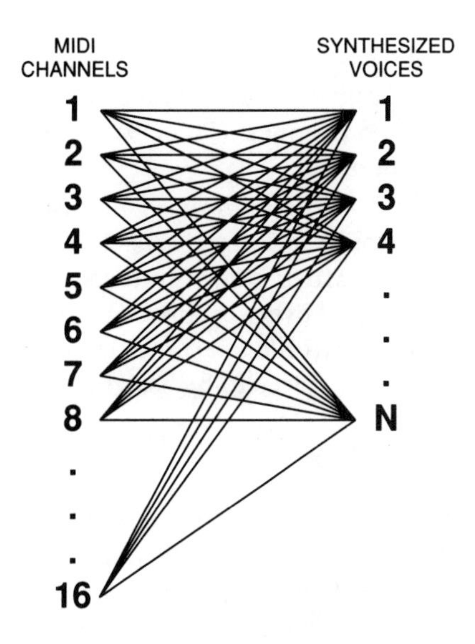

Fig. 2-7. Voice/channel assignment example of Mode 1 — Omni on/poly.

- *Mode 2 — Omni on/mono*: An instrument will assign any received note events to one monophonic voice, regardless of which MIDI channel it is received on (Fig. 2-8). This mode is rarely used.

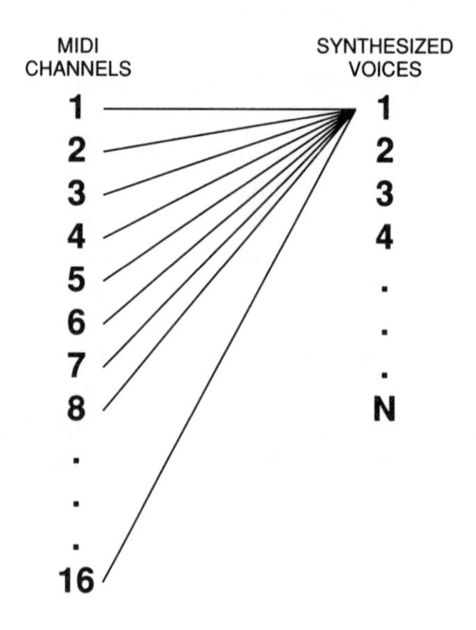

Fig. 2-8. Voice/channel assignment example of Mode 2 — Omni on/mono.

- *Mode 3 — Omni off/poly*: An instrument will be able to respond polyphonically to performance data that is transmitted over one or more assigned channels (Fig. 2-9). As such, it is commonly used by polytimbral instruments.
- *Mode 4 — Omni off/mono*: An instrument will only be able to generate one MIDI note per channel (Fig. 2-10). A practical example of this mode is often used in MIDI guitar systems, whereby MIDI data is monophonically transmitted over six consecutive channels (one channel/voice per string). Other electronic instruments may make use of this mode and may allow individual voices to be assigned to any combination of MIDI channels.

Channel-Voice Messages

Channel-voice messages are used to transmit real-time performance data throughout a connected MIDI system. They are generated whenever the controller of a MIDI instrument is played, selected, or varied by the performer. Examples of such control changes would be the playing of a keyboard, program selection buttons, or movement of modulation or pitch wheels. Each channel-voice message contains a MIDI channel number within its status byte and is able to be

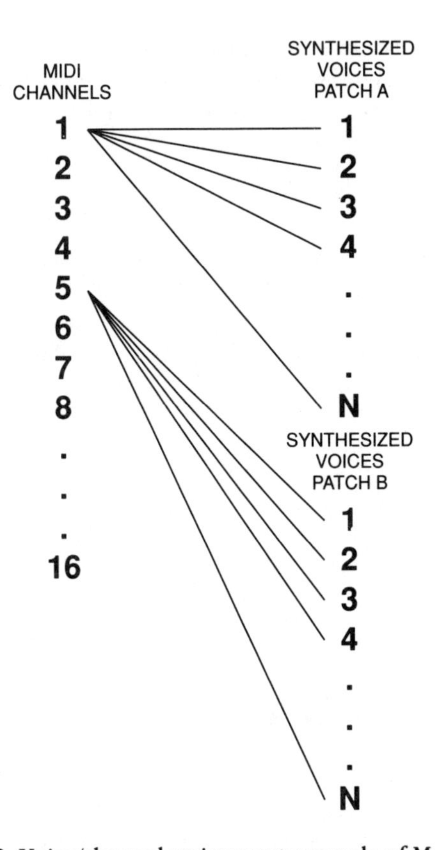

Fig. 2-9. Voice/channel assignment example of Mode 3 — Omni off/poly.

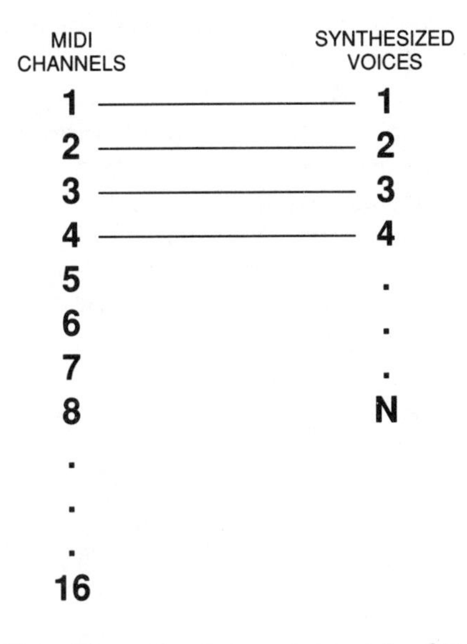

Fig. 2-10. Voice/channel assignment example of Mode 4 — Omni off/mono.

responded to by a device that is assigned to the same channel number. There are seven channel-voice message types: note-on, note-off, polyphonic-key pressure, channel pressure, program change, control change, and pitch-bend change.

Note On

A *note-on message* is used to indicate the beginning of a MIDI note. It is generated each time a note is triggered upon a keyboard, drum machine, or other MIDI instrument (by pressing a key, drum pad, or playing a sequence).

A note-on message consists of three bytes of information (Fig. 2-11): a note-on status/MIDI channel number, MIDI pitch number, and attack velocity value.

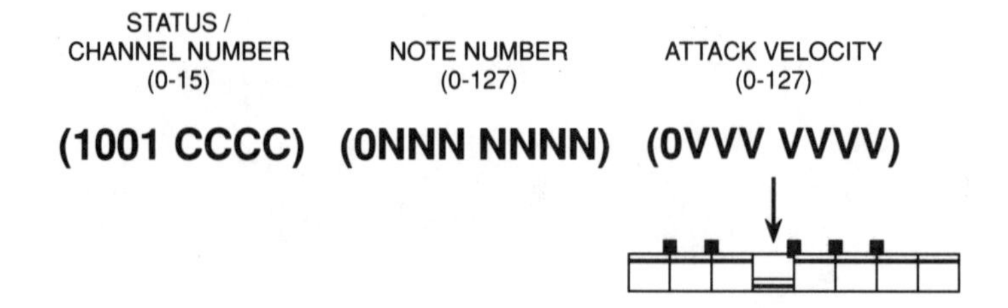

Fig. 2-11. Byte structure of a MIDI note-on message.

The first byte within the message specifies a note-on event and a MIDI channel (1–16). The second byte is used to specify which of the possible 128 pitches (numbered from 0–127) will be sounded by a MIDI instrument. In general, MIDI note number 60 is assigned to the middle C key of an equally tempered keyboard, while notes 21–108 correspond to the 88 keys of an extended keyboard controller.

The final byte is used to indicate the velocity or speed at which the key was pressed. The range of this value commonly varies from 1–127. And, it is used to denote the loudness of a sounding note that increases in volume with higher velocity values. Not all instruments are designed to interpret the entire range of velocity values (as with certain drum machines), while others do not respond dynamically at all. Instruments that do not support velocity information will generally transmit an attack velocity value of 64 for every note that is played, regardless of the velocity that is actually being played. Similarly, instruments which do not respond to velocity messages will interpret all MIDI velocities as having a value of 64.

A note-on message which contains an attack velocity of 0 (zero) is generally equivalent to the transmission of a note-off message. This indicates that the device is to silence a currently sounding note by playing it with a velocity (volume) level of 0.

Note Off

A *note-off message* is used as a command to stop playing a specific MIDI note. Each note that has been played via a note-on message is sustained until a corresponding note-off message is received. In this way, a musical performance can be encoded as a series of MIDI note-on and note-off messages. It should also be pointed out that a note-off message will not cut off a sound, it will merely stop playing it. If the patch being played has a release (or final decay) stage, it will begin that stage upon receiving this message.

As with the note-on message, the note-off's structure consists of three bytes of information (Fig. 2-12): a note-off status/MIDI channel number, MIDI pitch number (corresponding to the same numbers that are addressed by the note-on message), and release velocity value.

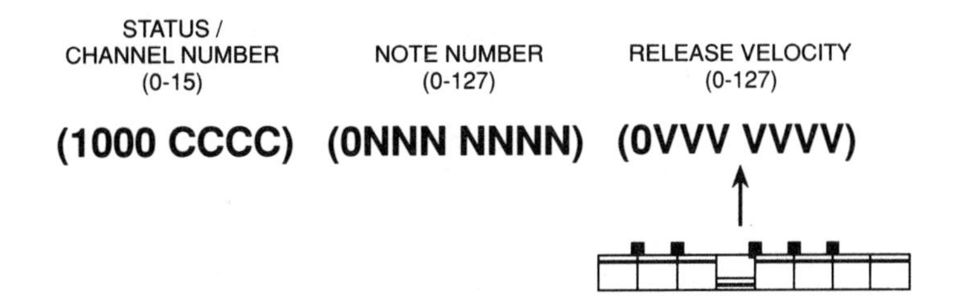

Fig. 2-12. Byte structure of a MIDI note-off message.

Conversely to the dynamics of attack velocity, the release velocity value (0–127) indicates the velocity or speed at which the key was released. A low value indicates that the key was released very slowly, while a high value shows that the key was released quickly. Although few instruments generate or respond to MIDI release velocity, instruments that are capable of responding to these values can be programmed so that a note's speed of decay is varied, often reducing the signal's decay time as the release velocity value is increased.

Polyphonic-Key Pressure

Polyphonic-key pressure messages are commonly transmitted by instruments that are capable of responding to the pressure changes which are applied to the individual keys of a keyboard. Such an instrument can be used to transmit individual pressure messages for each key that is depressed (Fig. 2-13).

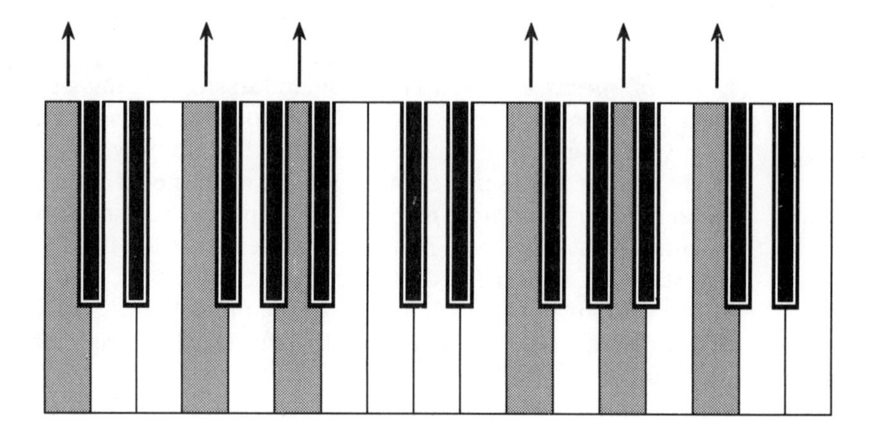

Fig. 2-13. Individual polyphonic-key pressure messages are generated when additional pressure is applied to each key that is played.

A polyphonic-key pressure message consists of three bytes of information (Fig. 2-14): the polyphonic-key pressure status/MIDI channel number, MIDI pitch number, and pressure value.

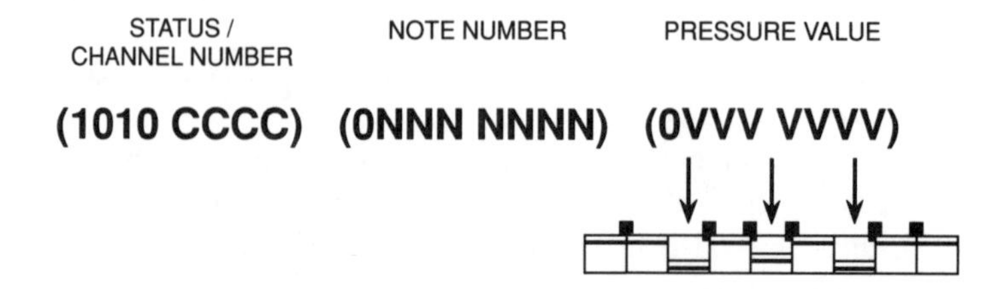

Fig. 2-14. Byte structure of a MIDI polyphonic-key pressure message.

The means by which a device will respond to these messages will often vary between manufacturers. However, pressure values may commonly be assigned to such performance parameters as vibrato, loudness, timbre, and pitch. Although controllers that are capable of producing polyphonic pressure are few and generally more expensive, it is not unusual for an instrument to respond to these messages.

Channel Pressure (After Touch)

Channel-pressure messages (often referred to as *after touch*) are commonly transmitted by instruments that will only respond to a single, overall pressure applied to their controllers, regardless of the number of keys being played at one time (Fig. 2-15). For example, if six notes are played upon a keyboard controller

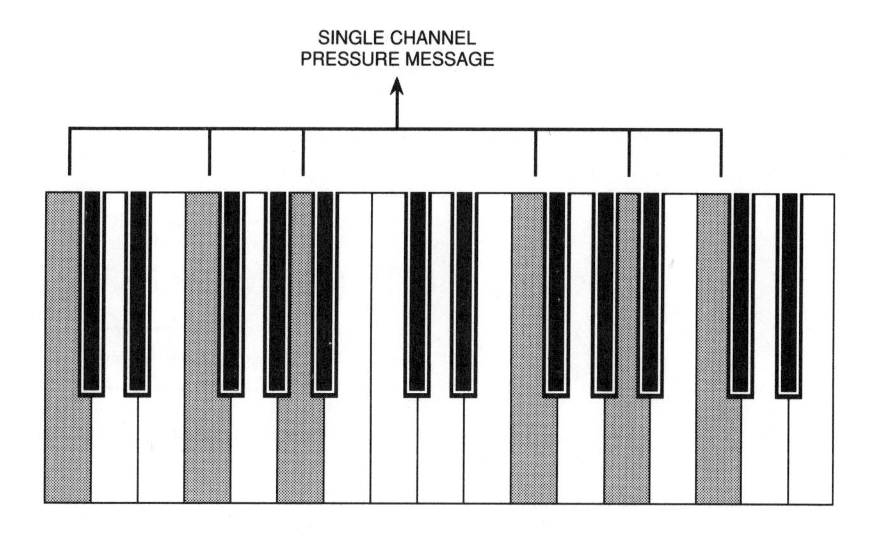

Fig. 2-15. One channel-pressure message effects all sustained notes that are transmitted over each MIDI channel.

and additional pressure is applied to only one key, all six notes would be affected. A channel-pressure message consists of three bytes of information (Fig. 2-16): channel-pressure status/MIDI channel number, MIDI pitch number, and pressure value.

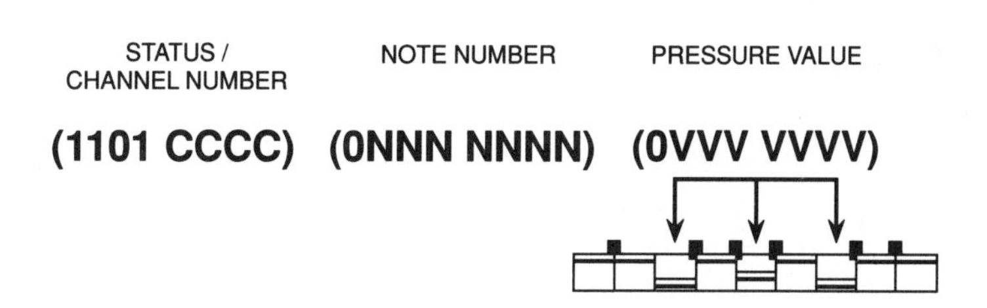

Fig. 2-16. Byte structure of a MIDI channel-pressure message.

As with polyphonic pressure changes, an instrument can often be programmed to respond to channel-pressure messages in many ways. For example, channel-pressure values may commonly be assigned to such performance parameters as vibrato, loudness, filter cutoff, and pitch.

Program Change

The *program-change message* is used to change the program or preset number that is active within a MIDI instrument or device. A *preset* is a user- or factory-defined number that will activate a specific sound patch or system setup. Up to 128 presets may be selected via MIDI using this message. A program-change message (Fig. 2-17) consists of two bytes of information: program-change status/ MIDI channel number (1–16) and a program ID number (0–127).

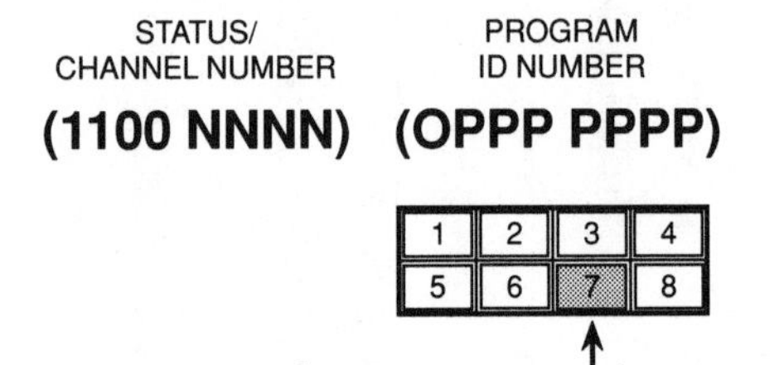

Fig. 2-17. Byte structure of a MIDI program-change message.

As an example, a program-change message can be used to switch between the various sound patches of a synthesizer from a remote keyboard instrument (Fig. 2-18). It can also be used to select rhythm patterns and/or setups within a drum machine, call up specific effects patches within an effects device, or a multitude of other controller and system setups that can be recalled as a program preset. Many MIDI devices allow for the recognition of these messages to be manually disabled and enabled by the user.

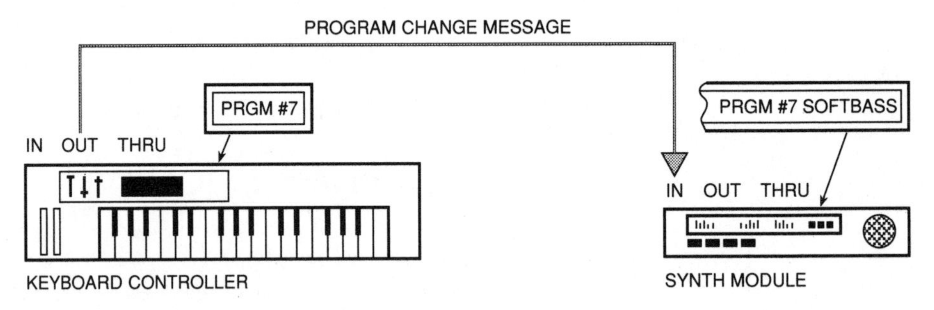

Fig. 2-18. Program-change messages may be used to change sound patches from a remote controller.

Control Change

The *control-change message* is used to transmit information that relates to real-time control over the performance parameters of a MIDI instrument. There are three types of real-time controls which can be communicated via control-change messages:

- *Continuous controllers*: Controllers that relay a full range sweep of possible control settings. Commonly, these range in value from 0–127.

However, two controller messages may be combined in tandem to achieve a greater resolution.

- *Switches*: Controllers that are either in an *on* or *off* state, with no intermediate settings.
- *Data controllers*: Controllers that enter data either through the use of a numerical keypad, or stepped up and down through the use of data-entry buttons.

A single control-change message or a stream of such messages is transmitted whenever controllers (such as foot switches, foot pedals, pitch-bend wheels, modulation wheels, breath controllers, etc.) are varied in real time. In this way, a controller may be used to correspondingly vary a wide range of possible parameters within an instrument or device in accordance with the controller's movements or commands. A control-change message (Fig. 2-19) consists of three bytes of information: the control-change status/MIDI channel number (1–16), a controller ID number (0–127), and corresponding controller value (0–127).

STATUS/ CHANNEL NUMBER	CONTROLLER ID NUMBER	CONTROLLER VALUE
(1011 NNNN)	**(0CCC CCCC)**	**(0VVV VVVV)**

Fig. 2-19. Byte structure of a MIDI control-change message.

Controller ID Numbers

The second byte of the control-change message is used to denote the *controller ID number*. This number is used to specify which of the device's program or performance parameters are to be addressed.

Although many manufacturers follow a general convention for assigning controller numbers to associated parameters, they are free to assign these parameters as they wish, provided they follow the defined format as provided by the MIDI specification (Fig. 2-20).

Controller Values

The third byte of the control-change message is used to denote the controller's actual data value. This value is used to specify the position, depth, or level of effect that the controller will have upon the parameter. In most cases, the value range of a 7-bit continuous controller will fall between 0 (minimum value) and 127 (maximum value) (Fig. 2-21). The value range of a switch controller is often 0 (off) and 127 (on) (Fig. 2-22A). However, switch functions are also capable of responding to continuous-controller messages by recognizing the values of 0–63 as off, and 64–127 as on (Fig. 2-22B).

14-BIT CONTROLLER MOST SIGNIFICANT BIT

Controller Number Hex	Decimal	Description
00H	0	Undefined
01H	1	Modulation Controller
02H	2	Breath Controller
03H	3	Undefined
04H	4	Foot Controller
05H	5	Portamento Time
06H	6	Data Entry MSB
07H	7	Main Volume
08H	8	Balance Controller
09H	9	Undefined
0AH	10	Pan Controller
0BH	11	Expression Controller
0CH	12	Undefined
.	.	.
0FH	15	Undefined
10H	16	General Purpose Controller #1
11H	17	General Purpose Controller #2
12H	18	General Purpose Controller #3
13H	19	General Purpose Controller #4
14H	20	Undefined
.	.	.
1FH	31	Undefined

14-BIT CONTROLLER LEAST SIGNIFICANT BIT

Controller Number Hex	Decimal	Description
20H	32	LSB Value for Controller 0
21H	33	LSB Value for Controller 1
22H	34	LSB Value for Controller 2
.	.	.
3EH	62	LSB Value for Controller 30
3FH	63	LSB Value for Controller 31

7-BIT CONTROLLERS

Controller Number Hex	Decimal	Description
40H	64	Damper Pedal (sustain)
41H	65	Portamento On/Off
42H	66	Sostenuto On/Off
43H	67	Soft Pedal
44H	68	Undefined
45H	69	Hold 2 On/Off
46H	70	Undefined
.	.	.
4FH	79	Undefined
50H	80	General Purpose Controller #5
51H	81	General Purpose Controller #6
52H	82	General Purpose Controller #7
53H	83	General Purpose Controller #8
54H	84	Undefined
.	.	.
5AH	90	Undefined
5BH	91	External Effects Depth
5CH	92	Tremolo Depth
5DH	93	Chorus Depth
5EH	94	Celeste (Detune) Depth
5FH	95	Phaser Depth

PARAMETER VALUE

Controller Number Hex	Decimal	Description
60H	96	Data Increment
61H	97	Data Decrement

PARAMETER SELECTION

Controller Number Hex	Decimal	Description
62H	98	Non-Registered Parameter Number LSB
63H	99	Non-Registered Parameter Number MSB
64H	100	Registered Parameter Number LSB
65H	101	Registered Parameter Number MSB

UNDEFINED CONTROLLERS

Controller Number Hex	Decimal	Description
66H	102	Undefined
.	.	.
78H	120	Undefined

RESERVED FOR CHANNEL MODE MESSAGES

Controller Number Hex	Decimal	Description
79H	121	Reset All Controllers
7AH	122	Local Control On/Off
7BH	123	All Notes Off
7CH	124	Omni Mode Off
7DH	125	Omni Mode On
7EH	126	Mono Mode On (Poly Mode Off)
7FH	127	Poly Mode On (Mono Mode Off)

Fig. 2-20. Listing of controller ID numbers, outlining both the defined format and conventional controller assignments.

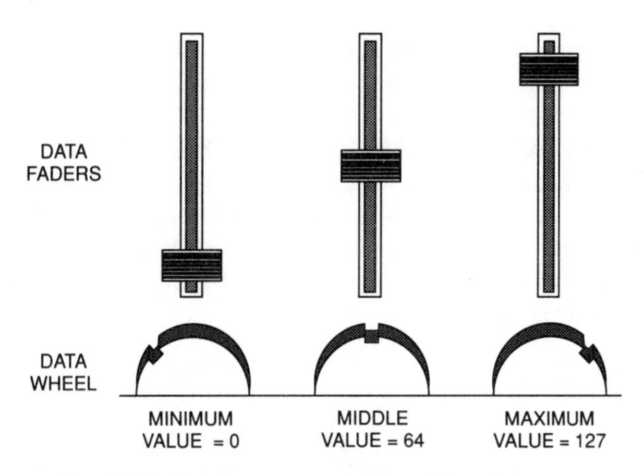

Fig. 2-21. Continuous-controller data value ranges.

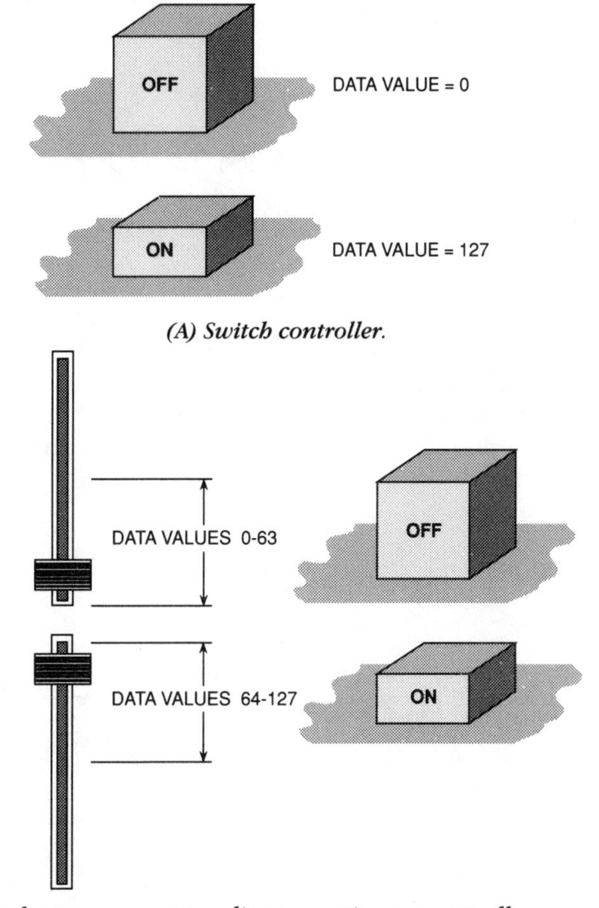

(A) Switch controller.

(B) A switch parameter responding to continuous-controller messages.

Fig. 2-22. Switch-controller data value ranges.

The practice of using the values of 0–127 to represent an increasing effect depth or signal level does not pertain to the control parameters of balance, panning, and expression.

A *balance controller* is used to vary the relative levels between two independent sound sources (Fig. 2-23). As with the balance control on a stereo preamplifier, this controller is used to set the relative left/right balance of a stereo sound source. The value range of this controller falls between 0 (full left sound source) and 127 (full right sound source), with a value of 64 representing a balanced stereo left/right field.

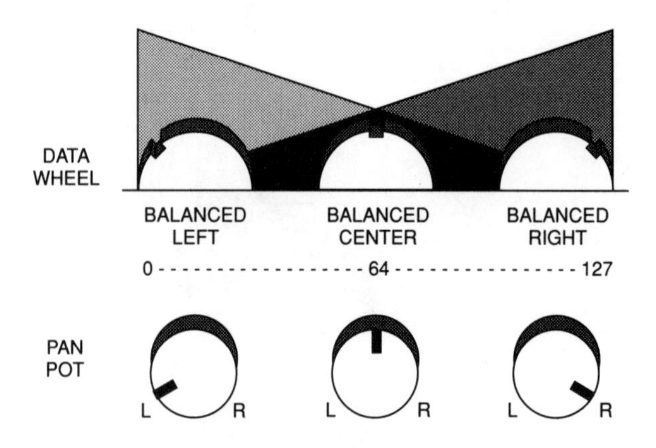

Fig. 2-23. Balance-controller data value ranges.

A *pan controller* is used to position the relative balance of a single sound source between the left and right channels of a stereo sound field (Fig. 2-24). The value range of this controller falls between 0 (hard left positioning) and 127 (hard right positioning), with a value of 64 representing a balanced center field.

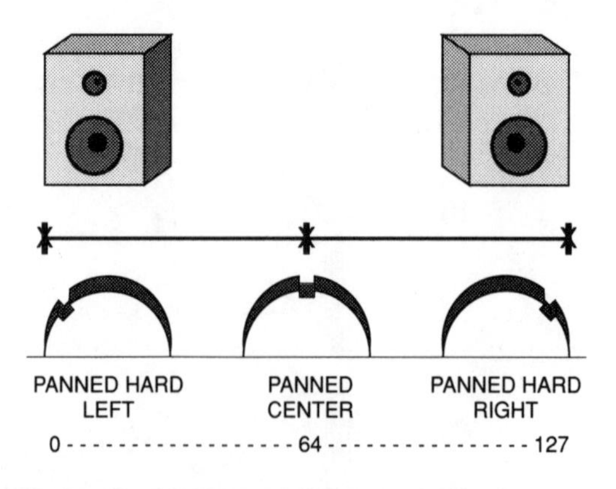

Fig. 2-24. Pan-controller data value ranges.

An *expression controller* is used to accent the existing level settings of a MIDI instrument or device. This control can be used to increase the channel volume level of an instrument, but cannot reduce this level below its programmed volume setting. The value range of this controller falls between 0 (current programmed volume setting) and 127 (full volume accent).

Controller ID Format

The following controller ID formats have been defined by the MIDI 1.0 Specification in order to ensure compatibility between manufacturers with respect to various controller parameters and their messages.

- *14-bit controllers*: Controller numbers 0–31 are commonly reserved for standard continuous-controller messages (such as modulation, breath, main volume, pan, etc.). However, should a greater resolution be required, it is possible to correspondingly link these messages with message numbers 32–63. Thus, instead of transmitting two data bytes over one message, four bytes will be transmitted over two messages. This will raise the resolution from a total of 127 possible steps to 16,683 steps.
- *7-bit controllers*: Controller numbers 64–95 are commonly reserved for switching functions (such as damper pedal, portamento, and sustain), or those related to the depth of a programmed effect (such as tremolo, chorus, and phase). As additional messages are not reserved for these functions their value range is limited to 0–127.
- *General purpose controllers*: These messages are not assigned to any particular controller function and may be assigned to any device-specific controller at the manufacturer's discretion.
- *Undefined controllers*: Undefined messages are currently not assigned to any particular controller function. However, they are reserved for future controller parameters.
- *Data-increment controllers*: These messages are used to transmit increment (+) and decrement (–) data to related controls.
- *Registered and nonregistered controllers*: Only three types of registered and nonregistered controllers are currently assigned to a controller function. These are: pitch-bend sensitivity, fine tuning, and coarse tuning.

Pitch-bend sensitivity refers to the response sensitivity (in semitones) of a pitch-bend wheel. The 7-bit value range of this controller permits adjustments in increments of up to 1/128 of the pitch-bend range. Fine tuning permits controlled adjustments (either up or down) by dividing a semitone into 8,192 possible pitch steps. Course tuning provides a maximum tuning range of up to 63 semitones above, and 64 semitones below the standard tuning pitch.

Pitch-Bend Change

Pitch-bend change messages are transmitted by an instrument whenever its pitch-bend wheel (Fig. 2-25) is moved either in the positive (raise pitch) or negative (lower pitch) from its central (no-pitch bend) position. Pitch-bend messages consist of three bytes of information (Fig. 2-26): a MIDI channel number, most significant bit (MSB), and least significant bit (LSB) value. Given that this data is transmitted over two data bytes, this message has an overall 14-bit resolution.

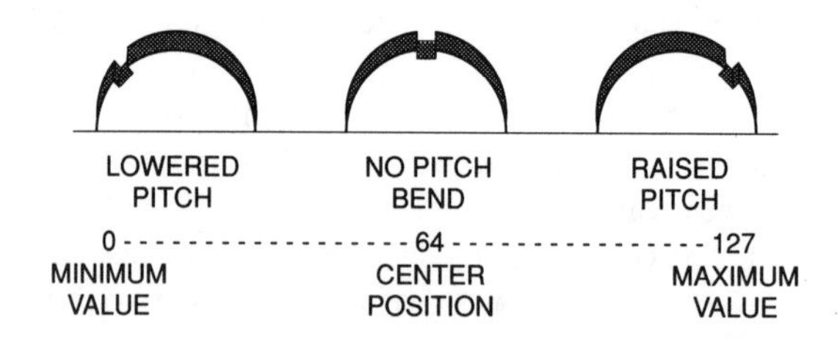

Fig. 2-25. Pitch-bend wheel data value ranges.

Fig. 2-26. Byte structure of a pitch-bend message.

Channel-Mode Messages

Controller numbers 121–127 are reserved for *channel-mode messages*. These include reset-all controllers, local control, all notes off, and MIDI mode messages.

Reset-All Controllers

This message is used to reinitialize all of the controllers (continuous, switch, and incremental) within one or more receiving MIDI instruments or devices to a standard, power-up default state.

Local Control

The *local-control message* is used to disconnect the controller of a MIDI instrument from its own voices (Fig. 2-27). This feature is useful for making a keyboard instrument a master performance controller within a MIDI system. Although an instrument's sound circuitry may be disconnected from its internal controller (when an instrument's local control is switched off), it is still capable of responding to MIDI performance and control messages from an external controller or sequencer.

A local-control message consists of two bytes of information: a MIDI channel number (1–16), and a local-control on/off status byte.

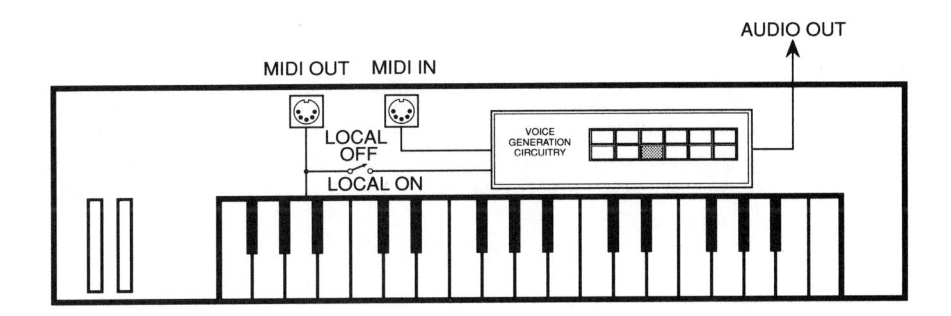

22757 Fig. 2-27

Fig. 2-27. Local-control on/off function.

All Notes Off

Occasionally, a note-on message will be received by a MIDI instrument, whereby the following note-off message is somehow ignored or not received. This unfortunate event often results in a stuck note that will continue to sound until a note-off message is received for that pitch. As an alternative to searching for this note, an *all-notes-off message* can be transmitted, which effectively turns off all of the 128 notes. As not all MIDI instruments are capable of responding to this panic-button message, it should be transmitted as a last resort.

Omni Mode Off

Upon the reception of an *omni mode-off message*, a MIDI instrument or device will switch modes (or remain in the omni-off mode), so it will respond to individually assigned MIDI channels instead of responding to all MIDI channels at once.

Omni Mode On

Upon receiving an *omni mode-on message*, a MIDI instrument or device will switch modes (or remain in the omni-on mode), so it will respond to all MIDI channel messages, regardless of which channels these messages are being transmitted on.

Mono Mode On

Upon receiving a *mono mode-on message*, a MIDI instrument will assign individual voices to consecutive MIDI channels, starting from the lowest currently assigned or basic channel. That is, the instrument can play only one note per MIDI channel. However, it is capable of playing more than one monophonic channel at a time.

Poly Mode On

Upon the reception of a *poly mode-on message*, a MIDI instrument or device will switch modes (or remain in the poly on mode). This message allows an instrument to respond to MIDI channels polyphonically. In this way, a device is able to play more than one note at a time over a given channel or number of channels.

System Messages

As implied by its name, *system messages* are globally transmitted to every MIDI device in the MIDI chain. This is accomplished because MIDI channel numbers are not addressed within the byte structure of a system message. This fact means that any device will respond to these messages, regardless of what MIDI channel or channels the device is assigned to. There are three system message types: system-common messages, system real-time messages, and system-exclusive messages.

System-Common Messages

System-common messages are used to transmit MIDI time code, song position pointer, song select, tune request, and end-of-exclusive data throughout the MIDI system or the 16 channels of a specified MIDI port.

MTC Quarter Frame

MIDI time code (*MTC*) provides a cost-effective and easily implemented means for translating *SMPTE* (*Standardized Synchronization Time Code*) into an equivalent code which conforms to the MIDI 1.0 Specification. It allows for time-based code and commands to be distributed throughout the MIDI chain. *MTC quarter-frame messages* are transmitted and recognized by MIDI devices that are capable of understanding and executing MTC commands.

A grouping of eight quarter frames are required to denote a complete time-code address (in hours, minutes, seconds, and frames). For this reason, SMPTE time is updated every two frames. Each quarter-frame message contains two bytes. The first is a quarter-frame common header. The second byte contains a nibble (four bits) which represents the message number (0 through 7), and a final nibble for each of the digits of a time field (hours, minutes, seconds, or frames). More in-depth coverage of MIDI time code can be found in Chapter 9.

Song Position Pointer

Song position pointer (*SPP*) permits a sequencer or drum machine to be synchronized to an external source (such as a tape machine) from any measure position within a song. The *SPP message* is used to reference a location point within a MIDI sequence (in measures) to a matching location upon an external device (such as a drum machine or tape recorder). This message provides a timing reference that increments once for every six MIDI clock messages, with respect to the beginning of a composition.

A SPP message is generally transmitted while the MIDI sequence has stopped, enabling MIDI devices equipped with SPP to chase in a fast forward motion through the song, and lock to the external source once relative sync is achieved. More in-depth coverage of the SPP can be found in Chapter 9.

Song Select

A *song-select message* is used to request a specific song from the internal sequence memory of a drum machine or sequencer (as identified by its song ID number). Upon being selected, the song will respond to MIDI start, stop, and continue messages.

Tune Request

This message is used to request a MIDI instrument initiate its internal tuning routine (if so equipped).

End of Exclusive

The transmission of an *end-of-exclusive (EOX) message* is used to indicate the end of a system-exclusive message. In-depth coverage of system-exclusive will be discussed later in this chapter.

System Real-Time Messages

The transmission of single byte *system real-time messages* provide the precise timing element required for synchronization by MIDI devices during performance. To avoid introducing timing delays, the MIDI specification allows system real-time messages to be inserted at any point within the data stream, even in the middle of other MIDI messages (Fig. 2-28).

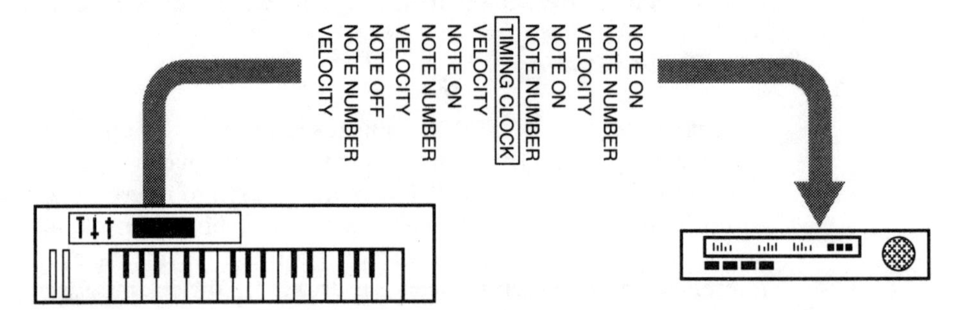

Fig. 2-28. System real-time messages may be inserted within the byte stream of other MIDI messages.

Timing Clock

The MIDI *timing-clock message* is transmitted within the MIDI data stream at a rate of 24 times per quarter note (24 ppqn). This message is used to synchronize the internal timing clocks of each MIDI device within the system and is transmitted in both the start and stop modes at the currently defined tempo rate.

Start

Upon the receipt of a timing-clock message, the MIDI *start command* instructs all connected MIDI devices to begin playing from the beginning of their internal sequence. Should a program be in midsequence, the start command will reposition the sequence back to its beginning, at which point it will begin to play.

Stop

Upon receipt of a MIDI *stop command*, all devices within the system will stop playing at their current position point.

Continue

After receiving a MIDI stop command, a MIDI *continue message* will instruct all connected devices to resume playing their internal sequences from the precise point at which it was stopped.

Active Sensing

When in the stop mode, an optional *active-sensing message* may be transmitted throughout the MIDI data stream every 300 milliseconds. This instructs devices able to recognize this message that it is still actively connected within the MIDI system.

System Reset

This message is manually transmitted in order to reset a MIDI device or instrument back to its initial power-up default settings (commonly Mode 1, local-control on, and all notes off).

System-Exclusive Messages

The *system-exclusive (SysEx) message* enables MIDI manufacturers, programmers, and designers to communicate customized MIDI messages between MIDI devices. It is the purpose of these messages to give manufacturers, programmers, and designers the freedom to communicate any device-specific data of an unrestricted length, as they see fit. Commonly, SysEx data is used for the bulk transmission and reception of program data, sample data, and real-time control over a device's parameters (as with editor/librarian programs).

The transmission format for SysEx messages (Fig. 2-29) as defined by the MIDI standard includes a SysEx status header, manufacturer's ID number, any number of SysEx data bytes, and an EOX byte. Upon receiving a SysEx message, the identification number is read by a MIDI device to determine whether or not the following messages are relevant. This is easily accomplished, as a unique one or three byte ID number is assigned to each registered MIDI manufacturer. If this number does not match the receiving MIDI device, the ensuing data bytes will be ignored. Once a valid stream of SysEx data is transmitted, a final EOX message is sent, after which the device will again respond to incoming MIDI performance messages.

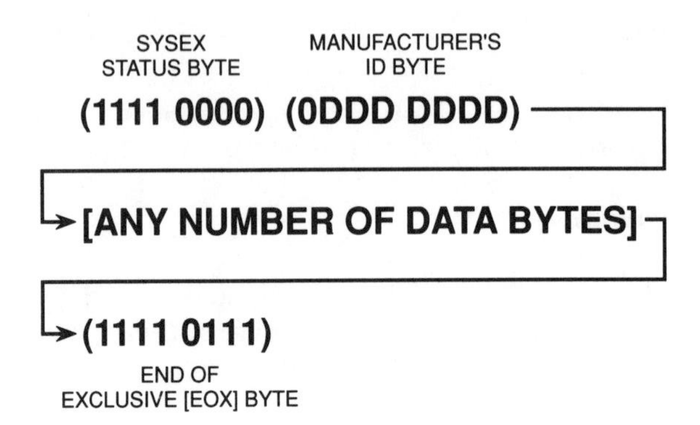

Fig. 2-29. System-exclusive data (one ID byte format).

Universal Nonreal-Time System Exclusive

Universal nonreal-time SysEx data is a protocol that is used to communicate control and nonreal-time performance data. It is currently used to intelligently communicate a data-handshaking protocol (informing a device of a specific state or event, as well as requesting specific data). It is also used for the transmission and reception of universal sample-dump data and finally to transmit MIDI time-code cueing messages. A universal nonreal-time SysEx message consists of four or five bytes that includes two sub-ID data bytes that identify which nonreal-time parameter is to be addressed. It is then followed by a stream of pertinent SysEx data.

Universal Real-Time System Exclusive

Currently, two universal real-time SysEx messages are defined. Both of them relate to the MTC synchronization code (which is discussed in detail within Chapter 9). These include full message data (relating to a SMPTE address) and user-bit data.

Running Status

Within the MIDI 1.0 Specification, special provisions have been made to reduce the need for conveying redundant MIDI data. This mode, known as ***running status***, allows a series of consecutive MIDI messages that have the same status byte type to be communicated without repeating redundant status bytes each time a MIDI message is sent. For example, we know that a standard MIDI message is made up of both a status byte and one or more data bytes. When using running status, however, a series of pitch-bend messages that have been generated by a controller would transmit an initial status and data-byte message, followed only by a series of related data (pitch-bend level) bytes, without the need for including redundant status bytes. The same could be said for note-on, note-off, or any other status message type. Upon receiving a message that includes a new status byte value, the device will respond to it in a standard fashion until the next series of like status bytes are encountered.

Although the transmission of running-status messages is optional, all MIDI devices must be able to identify and respond to this data-transmission mode.

MIDI Filtering

A *MIDI filter* is a dedicated digital device, on-board processor, or computer program that allows specific MIDI messages or range of messages within a data stream to be either recognized or ignored. A MIDI data filter can be thought of as a pass/no-pass digital switch that can be programmed to block the transmission of specific MIDI messages (Fig. 2-30), such as velocity on/off, program-change on/off, modulation on/off, SysEx on/off, etc.

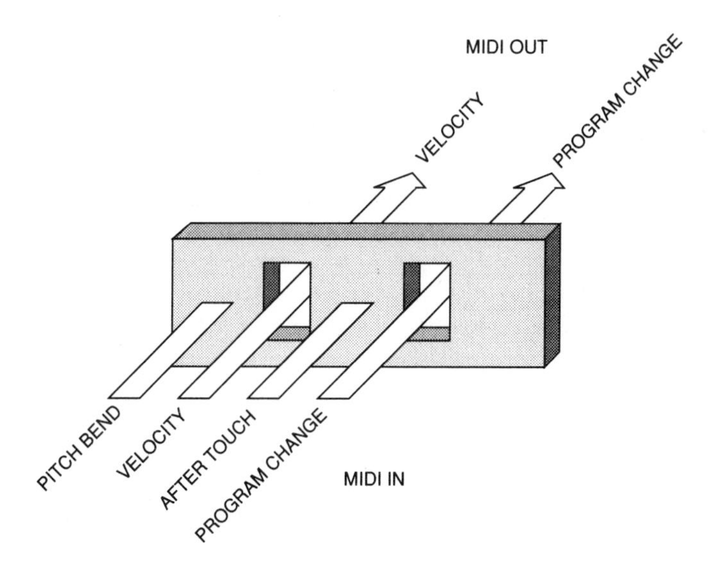

Fig. 2-30. A MIDI filter is used to block the transmission of specific MIDI messages.

A MIDI instrument or device may be capable of filtering incoming MIDI data or of filtering data at its MIDI out or thru ports. In the later case, the device itself will not be affected. However, all devices which follow in the chain may contain selectively filtered data. In addition, it is possible to filter messages that are transmitted upon user-selected MIDI channels (thus only affecting a specific device or instrument patch within the system).

MIDI Mapping

A *MIDI mapping device* is a dedicated digital system, on-board processor, or computer program that can be used to reassign the value of a data byte or range of data bytes to another user-defined value or range of values. Mapping may be applied within the MIDI chain in order to reassign channel numbers, program numbers, note numbers (for transposing notes or creating chords), controller numbers and values, etc. As with MIDI filtering, it is also possible to map specific message bytes that are transmitted upon user-selected MIDI channels.

Summary

From the information contained in this chapter, it can be seen that the current MIDI 1.0 Specification is a powerful digital medium for conveying both performance and control information throughout an electronic music system.

It may be felt by some that an in-depth coverage of MIDI messages will not strongly relate to them and their day-to-day production. It's true that in order to enjoy the full benefits of MIDI, a performer need not fully know the inner workings of the language. However, by gaining even a basic understanding of how MIDI devices communicate within a network, it is possible for us (as performers, hobbyists, designers, or programmers) to gain a better insight into the potential that is offered to us by the musical instrument digital interface.

Chapter 3

Hardware Systems within MIDI Production

In addition to the wide range of electronic MIDI instruments on the market, there is also a large number of MIDI hardware systems which have been designed to assist in distribution and MIDI processing. Such connection and hardware devices include systems which provide data management, processing, distribution, and interfacing. Since MIDI is a standardized language, its data can be communicated to all MIDI devices that are connected within the signal chain to integrate them into a single, efficient music production environment.

System Interconnection

As a data-transmission medium, MIDI is relatively unique within the world of sound production, allowing 16 channels of performance, controller, and timing data to be transmitted in one direction over a single MIDI cable. Using this medium, it is possible for MIDI messages to be communicated to a number of devices within a network from a single master controller (such as a keyboard or MIDI sequencer) over a single MIDI data chain. MIDI is also a flexible enough medium that any number of system interconnections can be made to allow for a wide range of possible system configurations.

The MIDI Cable

A MIDI cable (Fig. 3-1) consists of a shielded, twisted pair of conductor wires which has a male 5-pin DIN plug located at each of its ends. The MIDI specification makes use of only 3 of the possible 5 pins, with pins 4 and 5 being used as conductors for MIDI data, and pin 2 being used to connect the cable's shield to equipment ground. Pins 1 and 3 are currently not in use, but are reserved for possible changes in future MIDI applications. The twisted cable and metal mesh grounding are employed to reduce outside interference, such as radio frequency (RFI) or electrostatic interference, which can serve to distort or disrupt MIDI message transmission.

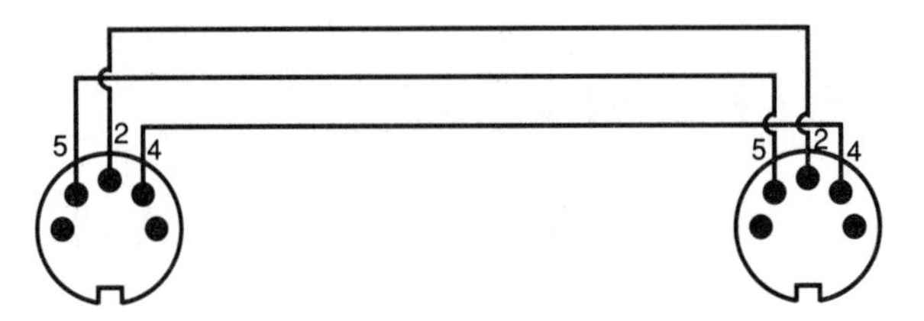

Fig. 3-1. Diagram of the MIDI cable.

Prefabricated MIDI cables in lengths of 2, 6, 10, 20, and 50 feet, may commonly be obtained from music stores that specialize in MIDI equipment. However, 50 feet is the maximum length specified by the MIDI specification to reduce the effects of signal degradation and external interference, which tends to occur over extended cable runs.

MIDI Ports

Within MIDI data distribution, there are three types of MIDI ports that make use of 5-pin DIN jacks to provide interconnections between MIDI devices within a network: MIDI in, MIDI out, and MIDI thru ports (Fig. 3-2). The hardware designs for these ports (as strictly defined by the MIDI 1.0 Specification) include the use of opto-isolators, which serve to eliminate possible ground loop problems within connected MIDI data lines.

MIDI In

The *MIDI in port* receives MIDI messages from an external source and communicates this performance, control, and timing data to the device's internal microprocessor.

More than one MIDI in port can be designed into a system to provide for MIDI merging functions, or for MIDI devices that can support more than 16 channels. Other devices (such as a MIDI controller) may not require the use of a MIDI in port at all.

MIDI Out

The *MIDI out port* is used to transmit MIDI messages from a single source to the microprocessor of another MIDI instrument or device.

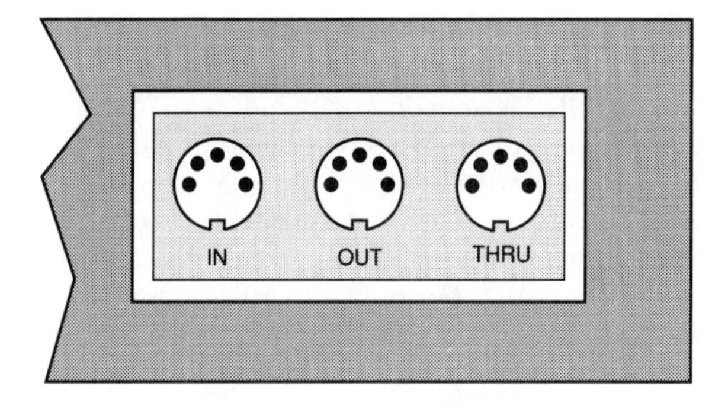

(A) MIDI ports on the rear of a MIDI device.

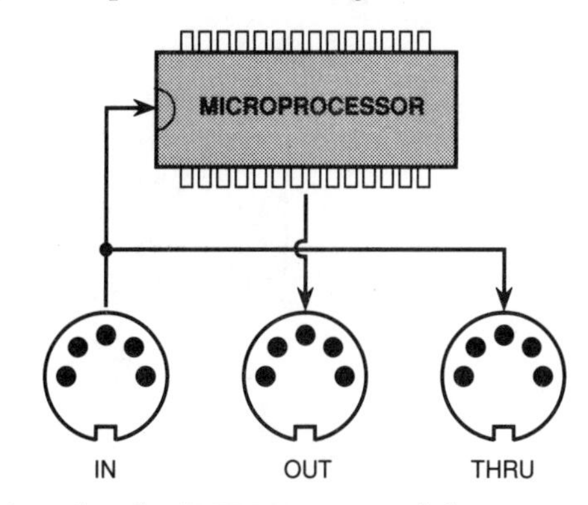

(B)Signal path of MIDI in, out, and thru ports.

Fig. 3-2. MIDI in, out, and thru ports.

More than one MIDI out port can be designed into a system for the simple purpose of providing multiple MIDI outs (providing distribution of the same data stream to a number of instruments). Alternatively, devices that can support more than 16 channels often have the ability to route individual MIDI data channel information to more than one isolated MIDI port. This has the advantage of providing a system with greater channel capabilities (providing data isolation between MIDI ports and reducing possible data clogging), and allowing the user to filter MIDI data on one port, while not selectively restricting the data flow within another port.

MIDI Thru

The *MIDI thru port* provides an exact copy of the incoming data at the MIDI in port. It is used to transmit this data out to another MIDI instrument or device that follows within the MIDI data chain. This port is used to relay an exact copy of the MIDI in data stream to the thru port, and is not merged with data that is transmitted at the MIDI out port.

MIDI Echo

Certain MIDI devices do not include a MIDI thru port. These devices, however, may offer a software-based transmission function for selecting between a MIDI out port and a *MIDI echo port* (Fig. 3-3). As with the MIDI thru port, the selectable MIDI echo function is used to provide an exact copy of any information that is received at the MIDI in port and route this data to the MIDI out/echo port. Unlike a dedicated MIDI out port, the MIDI echo function may often be selected to merge incoming data with performance data that is generated by the device itself. In this way, more than one controller can be simultaneously placed within a MIDI system. It is important to note that although performance and timing data is communicated by all devices that make use of MIDI echo, not all devices can echo SysEx data.

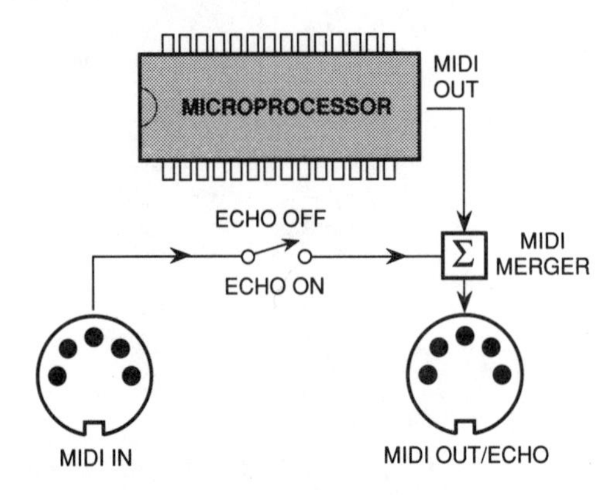

Fig. 3-3. MIDI echo configuration.

Typical Configurations

Although a wide range of device types and system designs exist from one MIDI setup to the next, there are a number of conventions that allow MIDI devices to be easily connected within any MIDI system. These common configurations allow MIDI data to be distributed in the most efficient manner possible.

As a primary rule, there are only two valid methods of connecting one MIDI device to another (Fig. 3-4):

1. Connecting the MIDI out port (or MIDI out/echo port) of one device to the MIDI in port of another device.
2. Connecting the MIDI thru port of one device to the MIDI in port of another device.

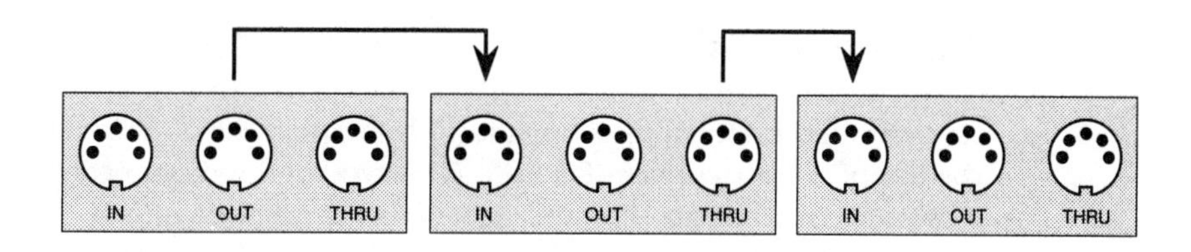

Fig. 3-4. The two valid means of connecting one MIDI device to another.

The Daisy Chain

One of the simplest and most commonly used methods for distributing data within a system is the ***MIDI daisy chain***. This method is used to distribute a single MIDI data line to every device within a system by transmitting data to the first device and subsequently passing an exact copy of this data through to each device within the chain (Fig. 3-5). This is done by sending the MIDI out data from the source device (controller, sequencer, etc.) to the MIDI in port of the second device. By connecting the MIDI thru port of the second device to the MIDI in port of the third device, this last device will receive an exact copy of the original source data at its input. This process may then continue throughout a basic MIDI system until the final device is reached.

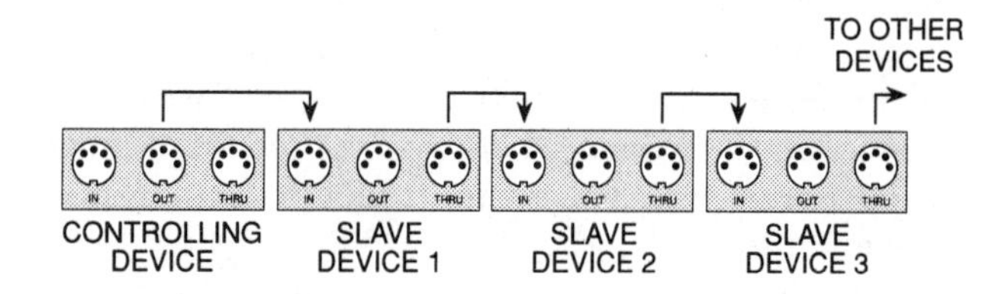

Fig. 3-5. Example of a connected MIDI system using a daisy chain.

As the MIDI thru port will only pass data that is an exact copy of the data at its MIDI in port, the signal can be traced back through each device to a single master controller (such as a keyboard). In most cases, this is acceptable because the controller is used to transmit data over one or more MIDI channels that will, in turn, be individually responded to by devices that have been assigned to these channels. Any device within a chain may be used as a controlling source by simply plugging its MIDI out port into the MIDI in port of any device that follows within the MIDI chain.

The Star Network

Another method for connecting multiple MIDI devices within a system is through the use of a ***star network*** (Fig. 3-6). This type of interconnection strategy allows a master controller to communicate with a number of MIDI instruments (or chains of instruments) over individually addressable MIDI ports. Data routing with such a network allows each set of MIDI in and out ports to accommodate independently isolated streams of MIDI data. This provides for the greatest flexibility in system interconnections.

Within larger, more complex MIDI systems, a star network offers a number of advantages over a daisy-chain network. It allows a controller, able to address each branch of a network, to communicate data (via MIDI in and out ports) that is relevant to each instrument or chain of instruments. As 16 channels of isolated MIDI data can be transmitted and received over each branch, this type of network is also ideal for use with MIDI devices that can handle more than 16 channels (e.g., a 32-channel sequencing program).

Midi patchbays (devices that are able to selectively route MIDI data paths) and newer generations of computer interfaces are often equipped with multiple MIDI in and out ports to facilitate the creation of a star network.

Wireless MIDI Transmission

A number of wireless MIDI systems are becoming available that allow a performer to transmit MIDI messages from a remote location to a stationary receiver which is connected within a MIDI system. Such a wireless system is useful for onstage applications, enabling a battery-operated MIDI controller (such as a MIDI guitar, wind, or remote keyboard controller) to be performed from any location in a live stage setting.

The TransMidi from MidiMAN (Fig. 3-7) is a device that allows most existing wireless systems to be converted into a MIDI wireless system. It consists of two modules: a receiver module and transmitter. The transmitter encodes standard MIDI data into an audio signal that can be transmitted using a standard wireless transmitter. This encoded audio signal is then received using a standard wireless receiver and is reconverted using the TransMidi receiver back into the original MIDI information.

Another such device is the MidiStar Pro from Gambatte, Inc. This all-digital wireless MIDI transmitter/receiver system is capable of transmitting MIDI data with a high degree of reliability at distances of up to 400 feet (120 meters). This data is transmitted using a patented UHF Spread Spectrum process that incorporates frequency diversity to transmit identical and reliable MIDI information over a wide range of radio frequencies.

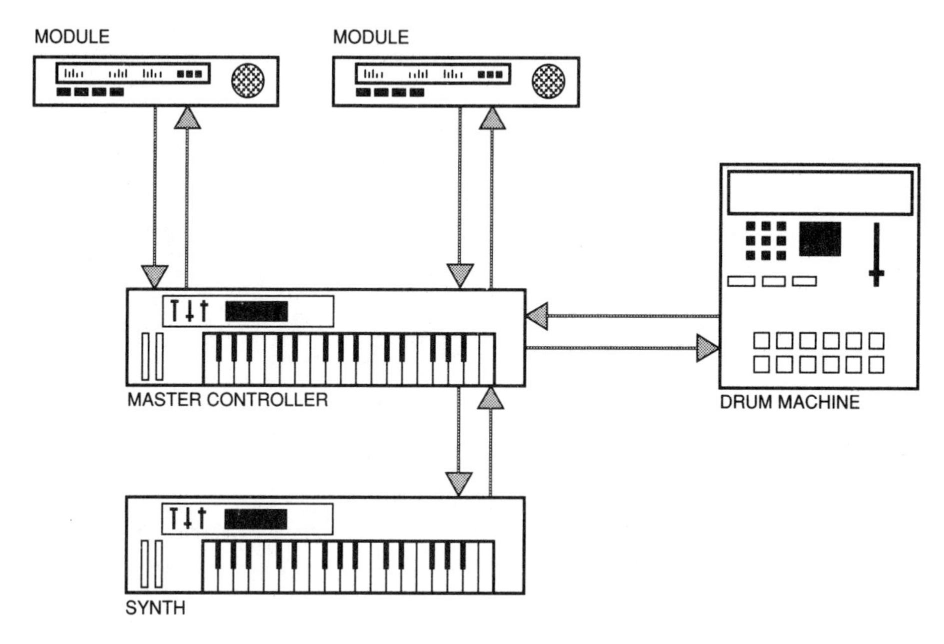

(A) Basic system interconnections.

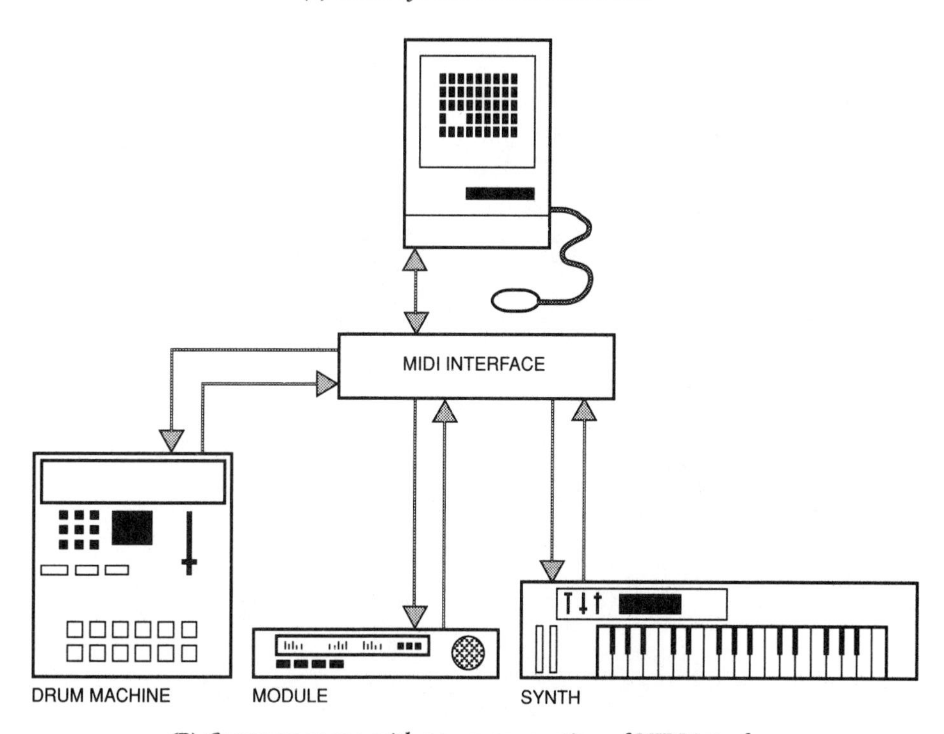

(B) Common usage with newer generation of MIDI interface.

Fig. 3-6. The star network.

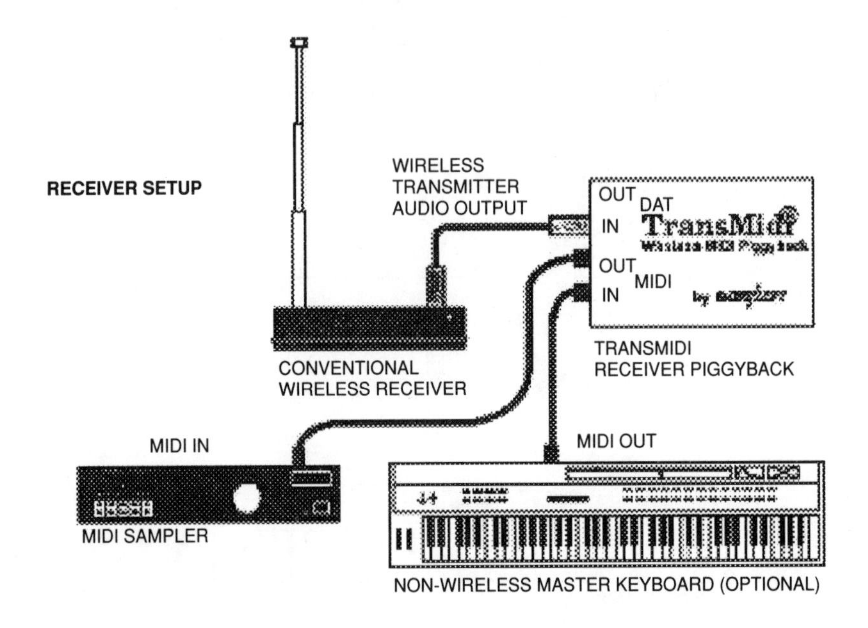

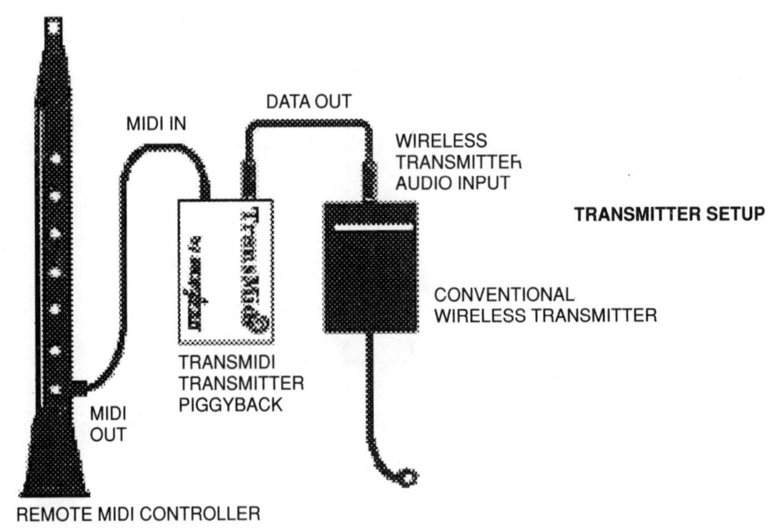

Fig. 3-7. The transMidi MIDI wireless transmitter and reciever systems. *(Courtesy of MidiMAN)*

The MidiStar Pro operates within the upper UHF band (902–928 megahertz [MHz]) to avoid interference with other wireless systems. Several factory-selected frequencies may also be selected, allowing multiple systems to be simultaneously operated.

Data-Management Hardware

There is a wide range of MIDI hardware devices on the market which are designed for the purpose of distributing and managing MIDI data within a system. Examples of these devices are the MIDI merger, MIDI thru box, MIDI switcher, MIDI patchbay, and MIDI data processors and diagnostic tools.

MIDI Merger

When combining the data from two or more separate MIDI lines into one MIDI data stream, a *MIDI merger* (Fig. 3-8) must be used. This must be done because it is not possible to simply splice two MIDI data lines into one input since the resulting data collisions would be irretrievably intertwined and not recognizable as valid MIDI data. Instead, it is the job of a MIDI merger to interleave the incoming data lines into a continuous MIDI stream that contains valid MIDI messages from each MIDI source.

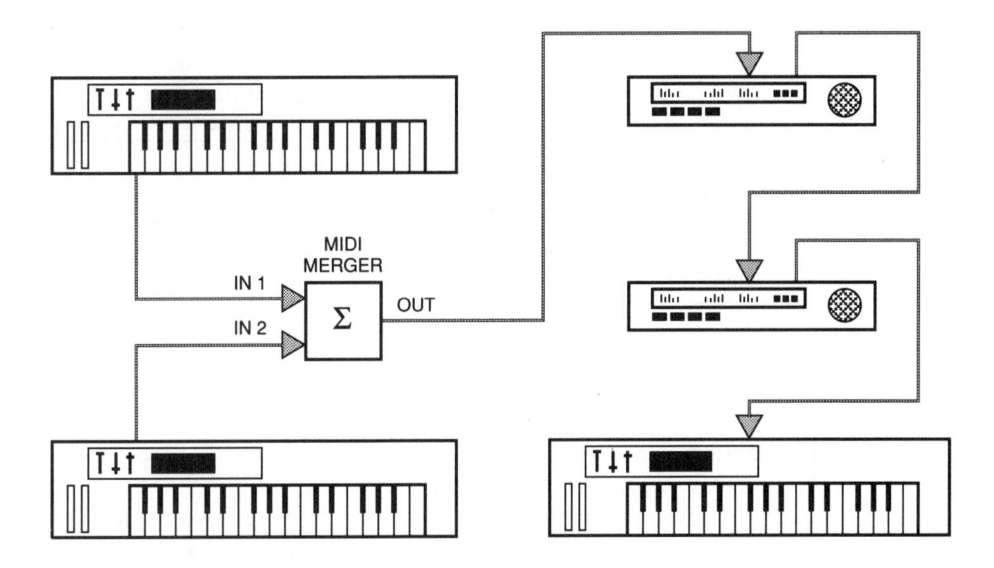

Fig. 3-8. Example of a MIDI merger within a system.

One function of a MIDI merging device would be to allow two MIDI devices to serve as master controlling sources within a MIDI system. For example, in Figure 3-8, the MIDI out data from both the keyboard controller or a MIDI wind controller are merged into one line so that either may be used to control an external MIDI device or system.

MIDI Thru Box

A *MIDI thru box* is used to distribute a MIDI data source throughout a system by providing an exact copy of a single incoming data signal to a number of MIDI thru ports (Fig. 3-9). Such a box can be used to replace the MIDI thru connections within a daisy-chained system, by effectively connecting each device to an exact copy of the source data. This is often done to avoid lags in transmission speed that occur within data chains that consist of four or more MIDI devices.

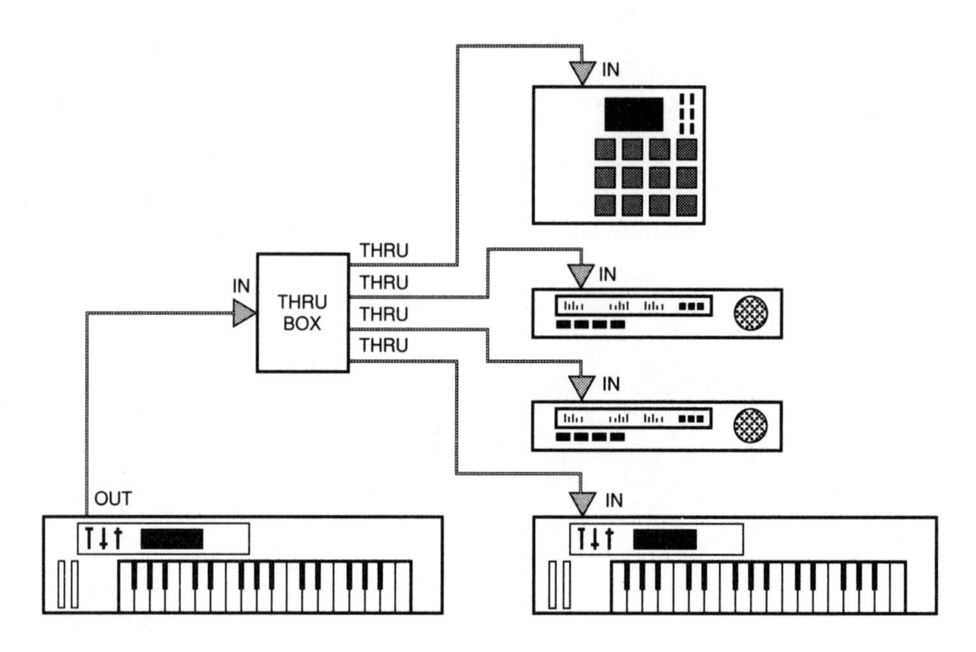

Fig. 3-9. Example of a MIDI thru box within a system.

MIDI Switcher

A *MIDI switcher* enables the user to select between two or more MIDI controller sources without the need for the manual repatching of MIDI cables. Figure 3-10 shows how a MIDI switcher might be used to select between a keyboard controller and a drum-pad controller within a system.

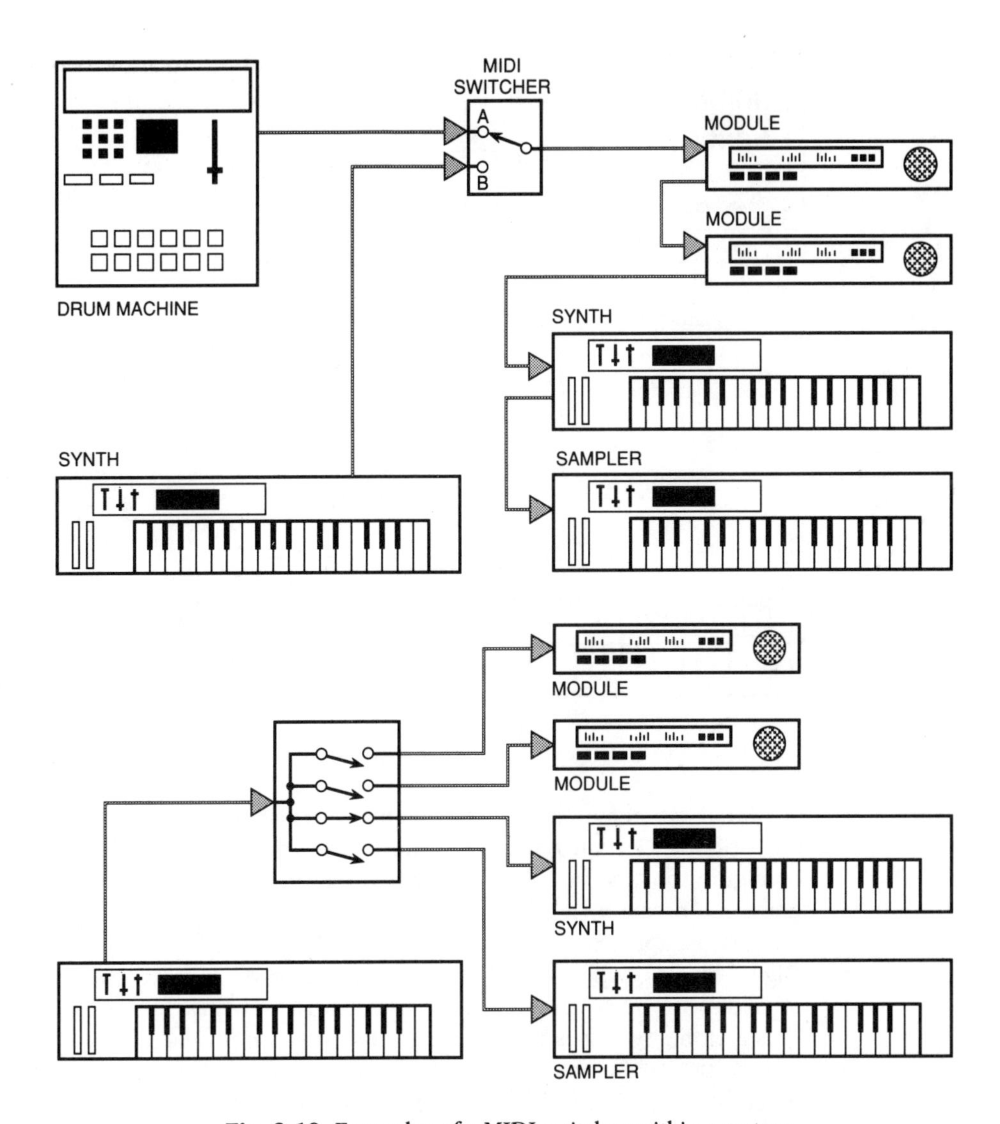

Fig. 3-10. Examples of a MIDI switcher within a system.

The MIDI Patchbay

Whenever a number of slaved MIDI devices are under the control of one or more MIDI controllers, it is often useful to employ the use of a *MIDI patchbay*. It is the function of a MIDI patchbay to selectively route MIDI data signal paths within a production system. Such a device can be used to route a single MIDI controller source to one or more MIDI devices within a connected system (Fig. 3-11). Alternatively, multiple controllers could be selected to simultaneously control separate MIDI devices within a system.

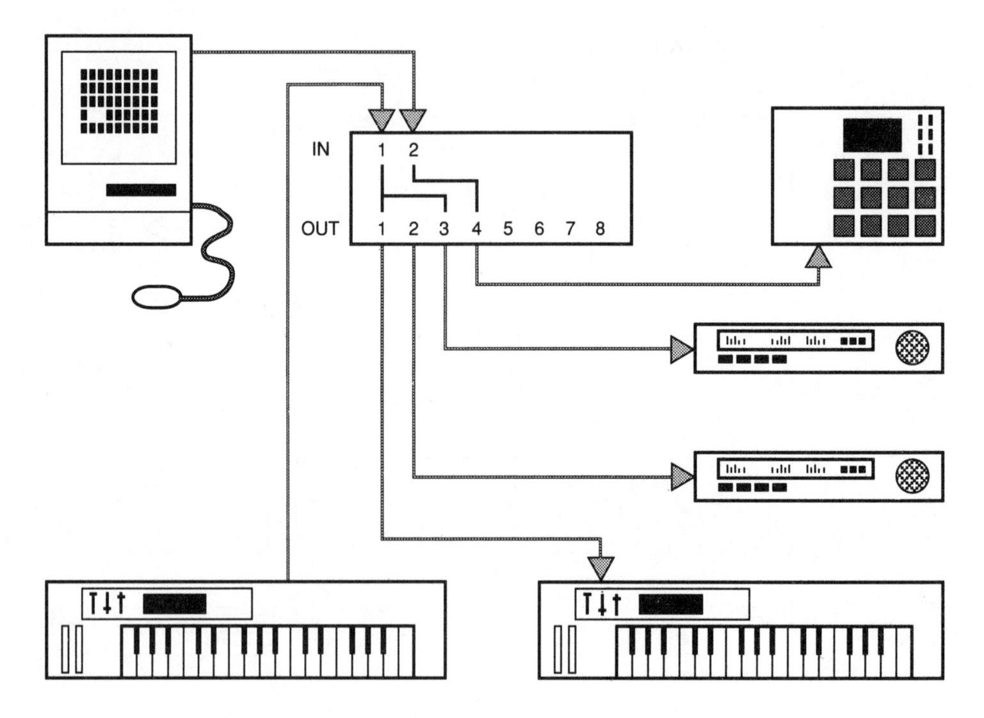

Fig. 3-11. Various examples of a MIDI patchbay within a system.

MIDI patchbay systems are available with a wide range of port configurations and options. Certain systems require that the user manually switch the system's configurations, while others allow the user to recall configurations using programmed preset locations.

One example of a basic system is the MX-28S 2 in 8 out MIDI patchbay from Digital Music Corp. (Fig. 3-12). This device allows the user to select the flow of MIDI data from two independent control devices to any eight MIDI slave devices. This is done by incorporating a 3-position switch for each output port used to select either of the possible inputs as a control source for each MIDI output port. When the switch is in the central off position, no data will be distributed to the selected output port.

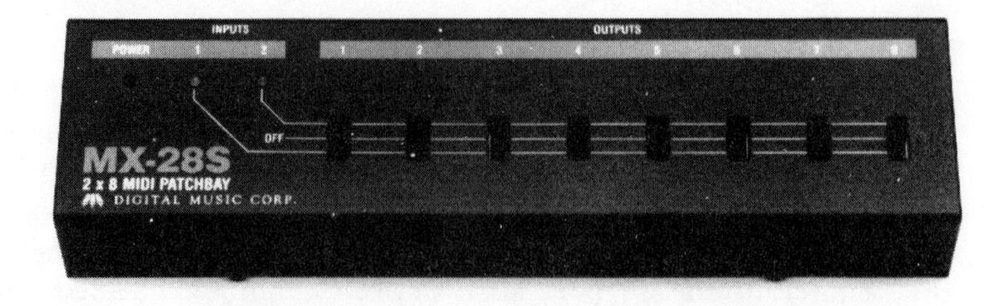

Fig. 3-12. The MX-28S MIDI patchbay. (*Courtesy of Digital Music Corp.*)

The MSB 16/20 from J.L. Cooper Electronics (Fig. 3-13) is a another MIDI patchbay device that is a 16 input by 20 output fully programmable MIDI switching system. This device is designed to allow the large studio or stage MIDI user to tie all MIDI sources and destinations into one box. It is capable of storing up to 64 different routing setups within a battery-backed memory that can be remotely recalled via MIDI program-change commands. Both the MIDI receiving input port and channel numbers are programmable and can be assigned to any individual output jack or merged into any output.

Fig. 3-13. MSB 16/20 programmable MIDI switching system. *(Courtesy of J.L. Cooper Electronics)*

MIDI Data Processors

Just as signal processors can be used within an audio chain to create an effect by changing or augmenting an existing sound, a *MIDI processor* can be placed within the MIDI data chain to route or alter MIDI messages within the system.

A MIDI processor is often capable of performing a wide range of algorithmic functions upon one or more MIDI signals, such as:

- MIDI channel and continuous-controller reassignments
- Programmable MIDI delays
- MIDI message filtering
- Velocity scaling and limiting
- Chromatic transposition
- Note, velocity, and pitch-wheel inversions

One example of a multifunction MIDI processing system is the MIDIbuddy multi-MIDI processor from ACME Digital Inc. This device features 10 independent MIDI in ports and 10 MIDI out ports (providing patchbay and routing control over a total of 160 channels), MIDI processing (including MIDI merging, delay, filtering, and mapping), SysEx data filing (for storing universal patch, and other device-specific data), and direct from disk sequence record/playback capabilities.

The MIDIbuddy allows the user to automate system patch and setup functions (via control messages or triggered in real time from a remote MIDI controller). In addition to a wide range of processing capabilities, this device is able to store MIDI sequences and SysEx data within one or two of the device's internal 3.5-inch floppy disk, or to an external hard disk/computer system (via an optional SCSI port).

MIDI Diagnostic Tools

Although a wide range of software-based tools exist for analyzing MIDI data through the use of a computer, a number of hardware tools are also available for diagnosing MIDI data within a MIDI production studio or live stage setting. Such tools are often used for detecting the presence of MIDI data or for trouble-shooting specific MIDI message types.

An example of a simple diagnostic tool is the MIDI Beacon from Musonix (Fig. 3-14). This pocket-sized device is used to indicate the presence of MIDI data within a system. This is accomplished through the use of an *LED* (*light emitting diode*) that flashes whenever MIDI data is present. This simple device can be used as a data indicator light for troubleshooting data within an interconnected chain or for checking MIDI cables.

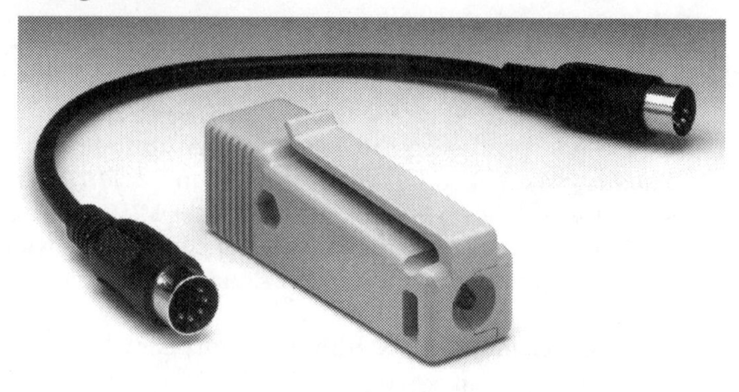

Fig. 3-14. The MIDI Beacon. (*Courtesy of Musonix, Ltd.*)

The MA36 MIDI analyzer from Studiomaster, Inc. (Fig. 3-15) is a hand-held tool that indicates the MIDI channel(s) being used within a network, and displays the type of MIDI messages that are being transmitted or received. MIDI in and thru connections allow this analyzer to be placed within the data chain and will remain active whenever the device is switched off.

The Local Area Network

When large volumes of channels, notes, SysEx, sample-dump data, and the like are crammed through a single serial data line at 31.25 kbaud within a large-scale MIDI production system, you're bound to run into a traffic jam which will often result in MIDI lag.

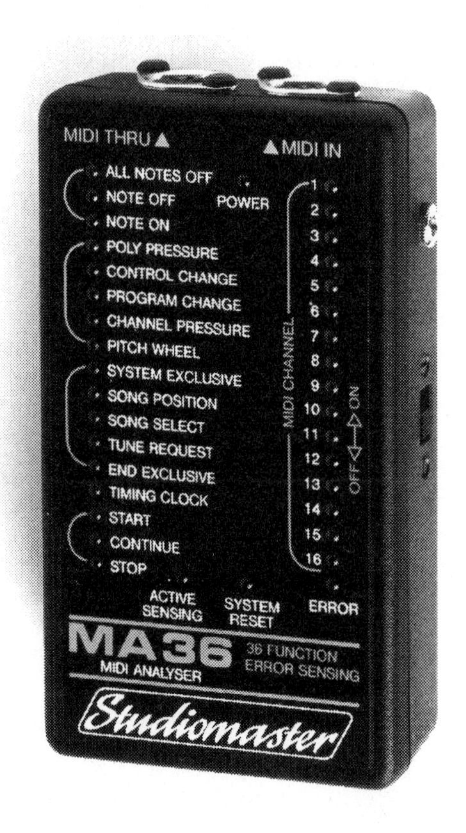

Fig. 3-15. The MA36 MIDI analyzer. (*Courtesy of Studiomaster, Inc.*)

A large part of these distribution and system configuration problems are currently being addressed by newer generations of MIDI interface, patch, and processing systems. However, one such MIDI management system is gaining recognition by tackling the problems of MIDI distribution and system setup through the use of a *local area network* (*LAN*) (Fig. 3-16), a technology widely used within the world of business. One of the major assets of a LAN is that it allows a multitude of separate computers (or in this case MIDI systems) to be directly connected, enabling them to exchange data in real time at a much higher transfer rate than is possible via MIDI.

The first of such LAN systems for use within MIDI production is the MidiTap from Lone Wolf. The MidiTap is a single-space rack device that has a front panel which includes four software control buttons, a small *LCD* (*liquid crystal display*) panel, and a velocity-sensitive parameter dial. Its rear panel features four MIDI in and out ports, an RS232/422 serial port for direct connection to any computer or modem, and a set of fiber optic ports. It is these fiber-optic ports that make this device so unique because they make use of Lone Wolf's proprietary MediaLink protocol.

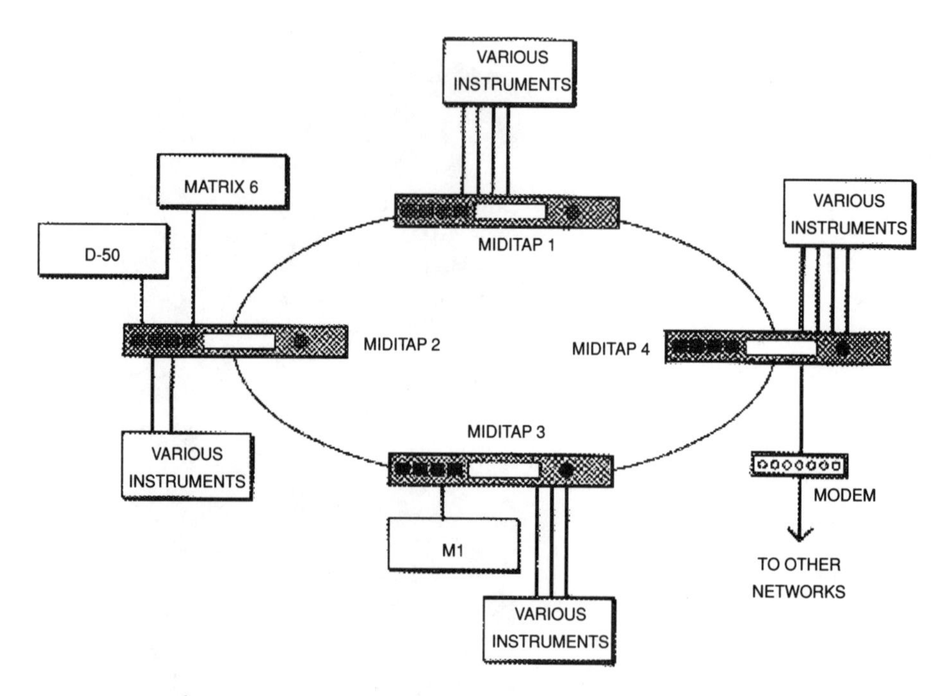

Fig. 3-16. Example of a local area network (LAN). (*Courtesy of Lone Wolf*)

MediaLink is a high-speed, bidirectional protocol that facilitates the efficient distribution of large amounts of digital information (such as MIDI, digital audio, SMPTE, etc.) in either real or nonreal time between connected MidiTaps or compatible devices. Unlike MIDI, which is limited to a transfer rate of 31.25 kbaud, MediaLink is capable of transmitting digital data in both directions at speeds of 1, 2, 4, 10, and 100 megabits per second. It is this high-data transfer rate that allows multiple MidiTaps to be linked together in a virtual LAN network, allowing any MIDI device connected to any MidiTap to be addressed by any other on the network in real-time performance.

In addition to its real-time performance mode, MidiTap also can be operated within a setup mode. This operating mode gives equal priority to all devices within the network, and it is used for data that does not need to be transmitted in real time (such as MIDI files, SysEx dumps, and other network configurations). Using this communications method, it is also possible to connect directly to another MidiTap via a telephone modem. In this way, such a remote system will be recognized as an active device within the network for direct up/downloading of system configurations, sound, or sample data.

In addition to real-time communications between devices and other LANs, the MidiTap is capable of creating an overall snapshot of a MIDI system, enabling a wide range of system parameters to be configured by the user and saved within the device or computer hard disk for later recall. Such a snapshot is defined by assigning various device setups within a system, each of which can be specified as being either a source device or a destination device. A device setup is more than a simple instrument identifier; it is a basic definition for assigning MIDI ports, channels, filters, and other parameters to one or more devices within the network.

The Personal Computer within MIDI Production

The personal computer is often a central component within most MIDI systems. Through the use of software programs and peripheral hardware, it is commonly used to process performance and setup related data from an integrated control position.

There are basically three personal computer types that are most commonly used within MIDI production: *Macintosh*, *IBM* (and IBM compatible), and the *Atari* family of computers. Each offers a wide range of computer systems, their own form and function, and their own distinct advantages and disadvantages to personal computing.

The major advantage to the personal computer is that it is a high-speed digital-processing engine that allows the user/artist to add peripheral options and hardware that best fit his or her production needs. It is available with a wide range of RAM memory options (by way of memory chip expansion, and floppy, hard, or optical disk options). One of the best advantages is the variety of software available for each computer system. It is this factor that turns the personal computer into a digital chameleon (i.e., able to change function in order to fit the necessary task at hand). In addition, the computer allows the user to choose software systems that best suits their production needs and personal approach to production.

Macintosh

A computer type that is widely accepted by music professionals is the Macintosh family of computers from Apple Computer, Inc. One of the major reasons for the success of the *Mac* is its graphic user interface which operates within a friendly window environment allowing multiple application windows to be moved, expanded, tiled, and stacked upon the system's monitor. This environment enables users to interface with the system using graphic icons and mouse-related commands.

The Macintosh SE (SCSI-enhanced) makes use of an 8-MHz Motorola 68000 processor, which provides the user with 1 megabyte (Mb) of RAM (expandable to 5 Mb), 1 expansion slot, keyboard, mouse, and 2 double-sided 800 kilobyte (kb) floppy disk drives (or one floppy and a hard disk). The Macintosh SE/30 (Fig. 3-17) is placed within the same housing as the SE, but makes use of a very fast Motorola 68030 processor and can process data much faster than the standard SE.

Fig. 3-17. The Macintosh SE/30 computer. (*Courtesy of Apple Computer, Inc.*)

The Macintosh II series from Apple Computer, Inc. (Fig. 3-18) uses a 68020 or 68030 processor, and provides the user with six hardware expansion slots (for added processing, graphic, video, and audio applications). The IIci offers speed, three expansion slots, and a 32-bit 68030 processor running at 25 MHz. Included within the IIci is an onboard 8-bit color-video chip that supports all Apple monitors (with the exception of the 2-page model). An 8-bit digital-to-analog converter is also included within most Macintosh systems, allowing a limited audio-quality signal to be reproduced directly from digitized computer files.

Fig. 3-18. The Macintosh IIci computer. (*Courtesy of Apple Computer, Inc.*)

The Macintosh portable personal computer offers complete compatibility with all Macintosh hard- and software in a portable design package. This system incorporates a 16-MHz 68000 processor, 640 x 400 pixel liquid crystal display, built-in keyboard, roller ball, battery, and disk storage. The portable is shipped in a standard 1-Mb configuration (expandable to 2 Mb), and is available with one or two built-in high-density Superdrives, or with both a Superdrive and 40-Mb hard disk.

IBM/Compatible

Due to its cost-effectiveness, acceptance, and sheer numbers within the business community, it is generally accepted that there are more IBM/compatible personal computers (Fig. 3-19) within the marketplace (musical or otherwise) than any other type. There is also a wide assortment of available IBM/compatible computers with a vast array of hardware and software peripherals for use with all forms of business and personal applications.

Fig. 3-19. Example of an IBM compatible computer. (*Courtesy of Cordata Technologies, Inc.*)

Although specific manufacturers and options are too great in number to mention here, there are four basic catagories of IBM/compatible systems. Each system makes use of various versions of MicroSoft's *disk operating system (DOS)* as a standardized operating system for performing basic management and processing tasks. The IBM PC and XT, and compatible computers employ standard 8- and 16-bit processing and generally operate at too slow a speed to offer enhanced graphic capabilities to the user. The IBM AT and 80286 compatible offer the user increased speed and access to greater amounts of memory. This system (which is based upon the Intel 80286 processor) may be run with the MicroSoft Windows operating system software, which offers a graphic window user-interface. Even faster systems for running MS-DOS based software (the standard software-based operating system for all IBM/compatible) have been developed using the 32-bit 80386 and 80486 processing chips. These systems are capable of accessing large amounts of memory and can be used for running several programs at a time (multitasking).

The standard XT or AT system will often offer a number of expansion slots which communicate with the computer via a 16-bit internal bus structure. These slots allow the user to add additional hardware application cards (such as video, mouse, scanners, FAX, MIDI interfaces, etc.) Certain 80386/80486-based systems make use of a 32-bit expansion slot for which few music-based hardware cards have been designed.

IBM/compatibles are commonly available in desktop, luggable, and lunchbox versions. In recent times, however, battery-operated IBM/compatible laptop computers have begun to play a prominent role within the PC marketplace. One laptop computer that has been specifically designed for use within music production is the Yamaha C1 Music Computer from Yamaha Corporation (Fig. 3-20). This fully IBM-compatible computer offers two MIDI in, one MIDI thru, and eight individually addressable output ports. A dedicated internal music timer is incorporated into the computer to enhance the accuracy of music software, along with an internal SMPTE time-code reader/generator, and built-in data sliders that can be used to directly input parameter or controller data. Music fonts and symbols are also embossed onto the typewriter-style keys for direct notation entry.

The C1 makes use of an 80286 processor and includes 1 Mb of memory (expandable to over 2.5 Mb). Standard computer accessories include external monitor port, printer, modem, mouse ports, and a Toshiba-style expansion slot for additional hardware support. This computer is available in two models: one with two 3.5-inch disk drive model, and a hard disk version that includes a 3.5-inch floppy and a 20 Mb hard-disk drive.

Atari

The Atari ST line of personal computers makes use of a strong graphic-interface environment and is also popular among musicians. It includes five models: 520 ST (Fig. 3-21), 1040 ST, Mega 2, Mega 4, and Stacy portable.

All five of these models are compatible, and nearly identical in design and operation. The major difference is the amount of memory that each computer is fitted with (512 kb, 1024 kb, 2048 kb, and 4096 kb consecutively). Any ST, however, can easily be upgraded to a RAM size of 4 Mb. Another major difference between the 520 (or 1040) and the Mega computing systems is that the later offers detachable keyboards, an expansion slot, and an expanded graphics processor.

Fig. 3-20. The Yamaha C1 music computer.
(Courtesy of Yamaha Corporation of America)

Fig. 3-21. The Atari 520 ST personal computer. *(Courtesy of Atari Corporation)*

The ST makes use of a window-based graphic-operating environment known as *GEM* (*Graphic Environment Manager*), which allows the user to interface with programs using drop-down menus, movable and resizable windows, icons, and desk accessories.

A cartridge port acts as an expansion slot by providing a direct access port for external hardware/software processing systems, such as ROM-based desktop accessories, hardware RAM systems, video systems, and digital audio systems.

The next generation Atari MIDI-compatible personal computer is the TT, a true 32-bit processor that will utilize the speed and power of the 68030 microprocessor (running at 16 MHz). This system includes a 2-Mb RAM as standard (expandable to 8 Mb), and is compatible with Atari's entire ST product line. The TT has all of the standard ports, plus 8-bit digital stereo sound, a floating point option, and six graphics modes.

Atari also offers two ST compatible laptop systems, known as the Stacy 2 and Stacy 4 (Fig. 3-22). These computers are consecutively configured with two or four megabytes of onboard memory, and include a 3.5-inch double-sided floppy-disk drive and a standard system speed of 8 MHz. For the increased memory applications, the Stacy 4 also includes a 40-Mb hard drive.

Fig. 3-22. The Stacy ST compatible laptop computer. (*Courtesy of Atari Corporation*)

The Stacy employs a Motorola 68C000 microprocessor which is compatible with all of the ST and Mega MIDI software packages, and includes MIDI ports as a standard feature. It also includes a back-lit super-twist LCD monochrome display with 640 x 400 resolution, making it easy to see under bright stage lights. For desktop use, the built-in monitor port allows the Stacy to be used with any ST monochrome or RGB color monitor. In addition to MIDI ports, the system includes all standard ST computer ports and interfaces, including monitor, serial, parallel, floppy disk, hard disk, R5232C, and game controller ports.

The MIDI Interface

Although both the MIDI protocol and the personal computer communicate via digital data, a digital hardware device (known as a *MIDI interface*) (Fig. 3-23) must be used to translate the serial message data of MIDI into a data structure that can be understood by the computer.

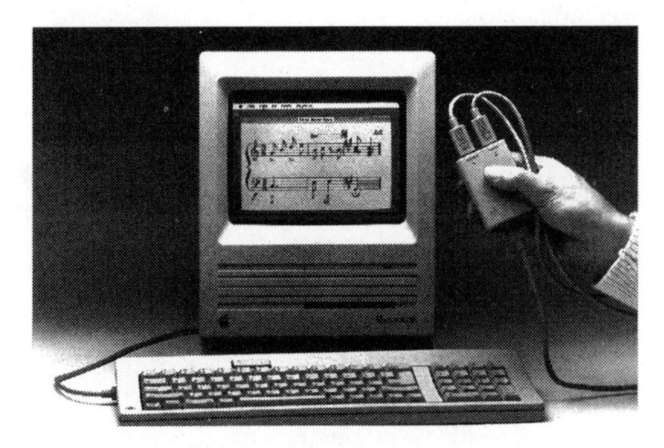

Fig. 3-23. The Macintosh SE with an Apple MIDI interface. (*Courtesy of Apple Computers, Inc.*)

A wide range of MIDI interfaces are available for use with most computer systems. Such devices range from passive systems (which essentially provide external MIDI ports and rely upon the computer to provide the conversion) to intelligent systems that incorporate internal processors for performing mundane calculations and function commands that would otherwise be left to the computer's internal processor.

Commonly, a MIDI interface will include facilities for maintaining synchronization between a multitrack tape machine and a MIDI sequencing program. Such a data-to-tape interface can be accomplished in a number of ways. *Frequency shift keying* (*FSK*) is one such sync method that encodes modulated data onto audio tape. Upon reproduction, this code (which is recorded upon a dedicated tape track) is then read and synchronized to an intelligent MIDI interface. Alternatively, similarly encoded data may be converted into SPP MIDI messages that can be used to maintain sync by driving the timing elements of other MIDI devices.

An example of a basic MIDI interface for the Macintosh computer is the MacNEXUS MIDI interface from J.L. Cooper. This passive device offers one MIDI in port, three MIDI out ports, input/output activity LEDs, and a Macintosh serial port connector. If more than 16 channels are needed, a second MacNEXUS can simply be attached to the second port.

The Studio 3 rack-mount MIDI and SMPTE interface from Opcode Systems, Inc. (Fig. 3-24) provides the Macintosh computer user with two independent MIDI in ports and six assignable MIDI out ports. Each MIDI out port can be assigned to either the printer or modem port. Front panel buttons allow *thru* patching for both serial ports, making it a simple task to switch on-line functions from MIDI to either the printer or modem (allowing the latter devices to share the same data lines with the MIDI interface, without the need for an external switching box).

Fig. 3-24. The Studio 3 MIDI and SMPTE interface. (*Courtesy of Opcode Systems, Inc.*)

The Studio 3 is capable of reading and writing all formats of SMPTE (a professional time-code format which is fully explained in Chapter 9) and is capable of converting SMPTE to either MTC or direct time lock. Jam sync capabilities are also implemented for regenerating SMPTE fresh time code when copying code from one tape machine to another. A Macintosh desk accessory is also included for external control over assignable MIDI outs and SMPTE.

The MIDI Time Piece from Mark of the Unicorn (Fig. 3-25), is a MIDI/SMPTE interface for the Macintosh family of computers that features eight independent MIDI inputs and eight outputs (providing up to 128 MIDI channels per time piece). This device offers complete MIDI merging, routing, channelizing, and event-muting capabilities. It can also function as a stand-alone merger/mapper when the computer is turned off.

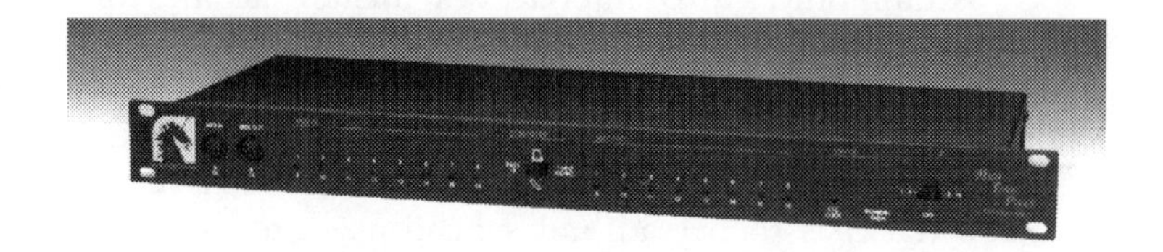

Fig. 3-25. The MIDI Time Piece multifunction MIDI interface. (*Courtesy of Mark of the Unicorn.*)

In addition to these functions, the time piece provides MIDI/SMPTE synchronization facilities for operation with tape and other external devices. This includes facilities for SMPTE sync (in all four formats), jam sync time-code regeneration, and conversion from SMPTE to either MIDI time code or direct time lock.

Up to four time pieces can be networked from either one or two Macintosh computers with a total channel capacity of up to 512 channels over 32 independent cables. Interconnecting network cables may range up to 1000 feet in length, allowing for multiple systems within a production facility, or long-run stage/booth MIDI interconnections. This device also includes a special fast mode for improved data management which serves to reduce MIDI data clog.

A Macintosh desk accessory is provided for the software, which allows control over all MIDI Time Piece control functions. These configurations may be accessed while another program is running and can be saved or loaded to computer disk memory. In addition, Mark of the Unicorn's sequencing program, Performer Version 3.4, allows the user to directly access and save MIDI Time Piece system configurations within a sequence file.

The MPU-401 MIDI interface from Roland (Fig. 3-26) was one of the first interfaces to be designed for the IBM XT/AT and compatible computers. Its programming and digital control structure has grown to be commonly regarded as the standard protocol, and it provides the basis for a multitude of other MPU-401 compatible interfaces.

This interface is made up of two system components: the 1/2-slot processing card and the interface box. The MPU-IMC features a MIDI in port, two MIDI out ports, sync out, tape in/out, and a metronome out jack. The sync out jack is used to synchronize external units (equipped with DIN sync) with the MPU's internal MIDI clock signal. The tape in/out jacks are used to transmit and receive FSK sync signals for synchronizing a sequence to recorded tape tracks. The metronome out jack plays a repeating tone for the recording of music within a steady metric timebase.

The MQX-32 professional PC MIDI interface from Music Quest, Inc. (Fig. 3-27), is an intelligent MPU-401 compatible device that provides the user with two separately controllable MIDI outputs. This allows supporting software to direct MIDI data to either (or both) MIDI out ports. If a 32 channel support is not used by the current software, the second MIDI out port will automatically echo the same data as the first port.

The MQX-32 is capable of reading and writing SPP data, which allows a supporting sequencer to synchronize to an external tape machine (in a chase/lock fashion) from any position within a song, without the need to rewind to its beginning point. SMPTE synchronization support is also included for reading and writing in the 30-drop and nondrop frame formats, in addition to fully supporting MTC.

This device is equipped with a short adapter cable that terminates in two standard MIDI out connectors and one MIDI in. RCA jacks extending through the card's mounting bracket provide tape in and out connections, in addition to an unpitched metronome output.

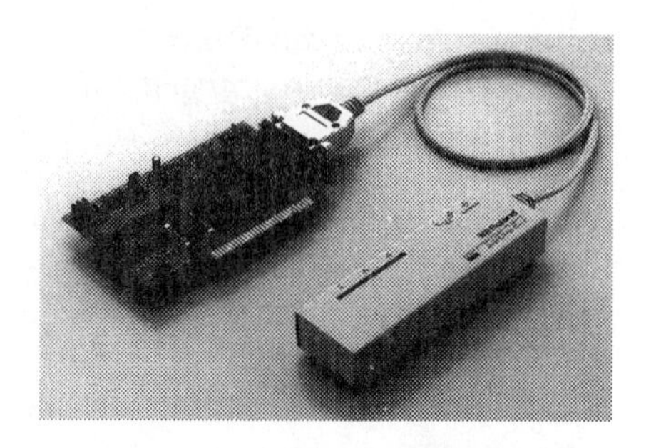

(A) System interconnections.

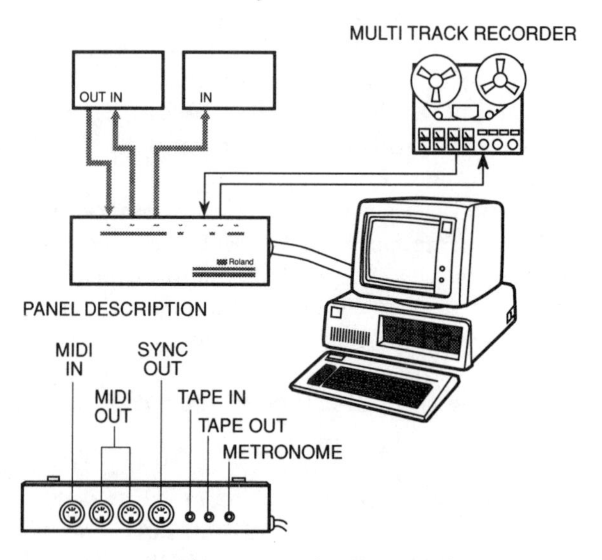

(B) The MPU-IPC-T MIDI interface.

Fig. 3-26. The MPU-401 MIDI interface. (*Courtesy of the Roland Corporation US*)

Unlike the Macintosh and most IBM/compatible computers, the Atari ST and Mega line of PC's, incorporate an internal MIDI interface as standard equipment, with MIDI in and out ports being housed directly within the computer. These ports can be directly addressed by programs running on the Atari, with data distribution being carried out by standard hardware distribution systems.

Computer-Based MIDI Applications and Utilities

One of the most recent advances within MIDI- and computer-based music applications is not taking place within the hardware environment at all, but within the area of software and operating systems for the personal computer. In

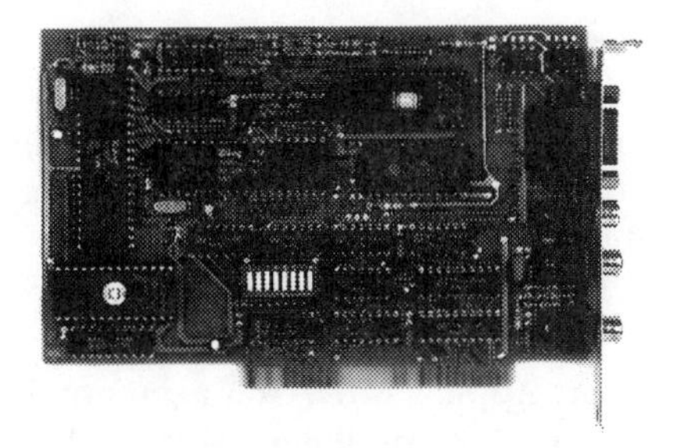

Fig. 3-27. The MQX-32 professional PC MIDI interface. (*Courtesy of Music Quest, Inc.*)

recent times, it has become increasingly common to find that hardware applications are being replaced by software equivalents. These advances within computer technology allow users to access these programs and applications more easily, in addition to their providing a means for processing tasks within a multiprogram environment.

Multitasking

A catalyst for a revolutionary advance within computer processing and electronic music production is the ability for many of the faster, more powerful PCs to process more than one program and/or task at a time. In actuality, the personal computer is not capable of processing multiple programs simultaneously, but instead switches between these programs at such a high rate of speed that these processing operations seem to blend together into one seamless structure. Such an environment is known as *multitasking*.

Concurrent program operation may be accomplished in varying degrees. For example, a process known as *program switching* allows the user to switch between two or more independently running programs. Programs, such as Juggler for the Atari ST and Microsoft Windows for the IBM, allow the user to instantly skip from one program to another without having to exit and reopen them. MultiFinder for the Apple Macintosh offers a basic form of multitasking by permitting certain operations (such as printing, disk backup, and file down/uploading) to occur in the background while primary applications are running in the foreground.

For computers whose operating systems permit true multitasking, the advantages are twofold. First, it is possible to gain control over and access to programs that are simultaneously open and running within the system. Secondly, direct communication could be set up between these programs or applications that would enable such programs to internally share data in real time.

Those within the recording and electronic music markets are commonly used to the idea of chaining one application device into another to perform a specific function. For example, a keyboard player might wish to filter out control-change messages, have access to the faders of a master controller, and record the final results into a sequencer. Under typical circumstances, a person would chain a filter into a master controller, and then into a sequencer. However, through the use of a computer that is working within a multitasking environment, it would be possible to internally chain together these programs and applications without the need for external hardware. This process, known as ***interapplication communication*** (*IAC*), is available for use with operating systems and MIDI programs that support Apple's MIDI Management Tools Set, Atari's MIDITasking (originally developed by Intelligent Music), Playroom Software's MIDI Executive for the IBM with Windows, and Mimetics SoundScape for the Amiga.

As such interapplication environments allow communication between internal computer applications, the MIDI output of one application must be internally routed to the MIDI input of another application, and finally to one or more MIDI out ports. This is accomplished through the use of a computer patchbay applications program.

A few practical examples of interapplication communication are currently being produced by Opcode Systems, Inc. For instance, their universal librarian package, Galaxy, is capable of exchanging patch-related data with their professional sequencing program, Vision. By running both programs at once, it is possible for patches that have been assigned or replaced within Galaxy to be automatically imported into the sequence files of Vision, while automatically updating any changes or deletions that relate to deleted patch files. Another popular example of IAC is Opcode's Studio Vision, which has come about as the result of a collaboration with Digidesign. Studio Vision is the combination of Opcode's professional MIDI sequencer, Vision, with the hard-disk audio recording capabilities of Digidesign's Sound Tools, whereby both program applications are simultaneously available to the user. Through the use of Studio Vision, it is possible to simultaneously reproduce a sequenced MIDI track along with two synchronized disk-based digital audio tracks (for reproducing sounds such as live vocals, sax, piano, or any other sound source) from a single Macintosh II or SE computer.

Chapter 4

The Electronic Musical Instrument

Since its inception in the early 1980s, the MIDI-based electronic musical instrument has become a central creative force within music technology and production. These devices, along with the advent of the multitrack home-recording market have made the personal music preproduction and production facility a strong reality of the 90s.

Although electronic instruments and their associated control devices will often vary in form and function, each makes direct use of a standard set of basic building-block components (Fig. 4-1) which include:

- *Central processing units (CPU)*: Small, dedicated computers (often in the form of a microprocessing chip) for interpreting a musician's actions and controlling the hardware that generates the audio-output signal.
- *Control panel*: Data-entry controls and display that allow patch and output functions to be manually changed by the user.
- *Performance controllers*: These include such controllers as keyboards, drum pads, wind controllers, etc. for inputting real-time performance data directly into the electronic instrument, or for transmitting this data into MIDI performance messages. Instruments without built-in controllers (commonly known as modules) respond to MIDI messages which are transmitted by external controlling instruments or devices.
- *Voice circuitry*: Allows for the generation and/or reproduction of analog audio signals, which can be amplified and heard via speakers or headphones.
- *Memory*: Memory is used for storing important internal data, such as patch information, setup configurations, and digital waveform data. This data may exist within a digital system as either *read-only memory* or *ROM* (memory which may only be retrieved from either factory-encoded memory chip, cartridge, or in certain applications, CD-ROM). Data may also be

stored within *random access memory* or *RAM* (memory that may be stored onto or retrieved from a memory chip, cartridge, hard disk, or optical disk).

- *Auxiliary controllers*: These are external controlling devices that are used in conjunction with a main instrument or controller. An example of such controllers are foot pedals (providing continuous-controller data), the breath controller (providing continuous breath-controlled data), the pitch-bending wheel, and modulation wheels. Additionally, certain controllers may be presented as a switched function for interpreting on/off information. An example of such controllers is the sustain pedal and vibrato control.
- *MIDI communications ports*: Provides the ability to transmit and/or receive MIDI data.

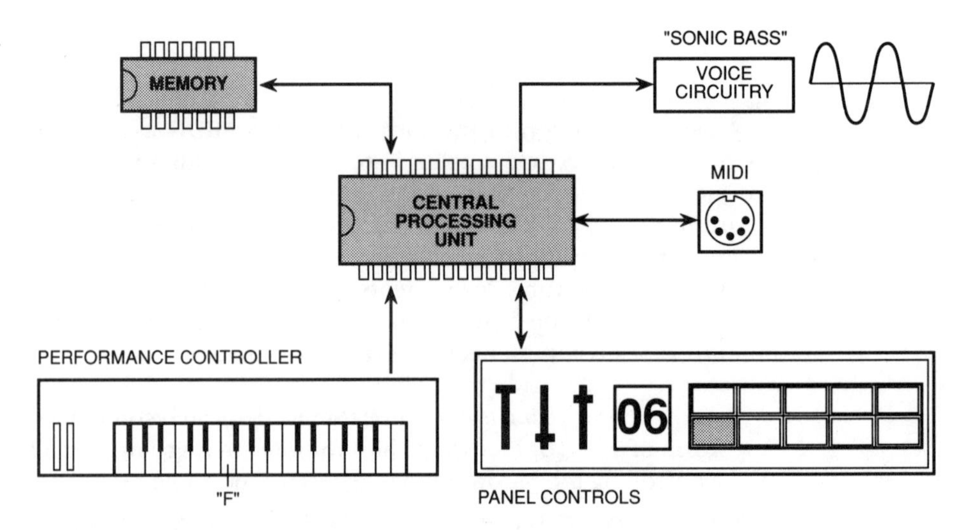

Fig. 4-1. The standard block components of an electronic musical instrument.

Although no direct internal communication link is made between each of these functional blocks, the data from each of these components is routed and processed through the instrument's CPU. For example, the user might wish to select a specific sound patch from the instrument's control panel. Upon pressing a patch selector button #6, instructions could be transmitted to the CPU to recall from memory all of the sound-patch parameters associated with a particular bass sound. These parameters are then used to shape the internal voice circuitry, so that when middle F on the keyboard is pressed, data is transmitted to the CPU, which, in turn, instructs the sound generators to output our bass patch at a frequency equivalent to the note value of middle F.

Real-time performance data is also commonly communicated between separate instruments or devices using the MIDI protocol (Fig. 4-2). For example, we can easily create a setup whereby the above keyboard/CPU combination would transmit note on/note off and other MIDI messages to another instrument which, in turn, will be instructed to output its internal voice circuitry (i.e., *soft celesta*) at a frequency that is equivalent to the note value of middle F.

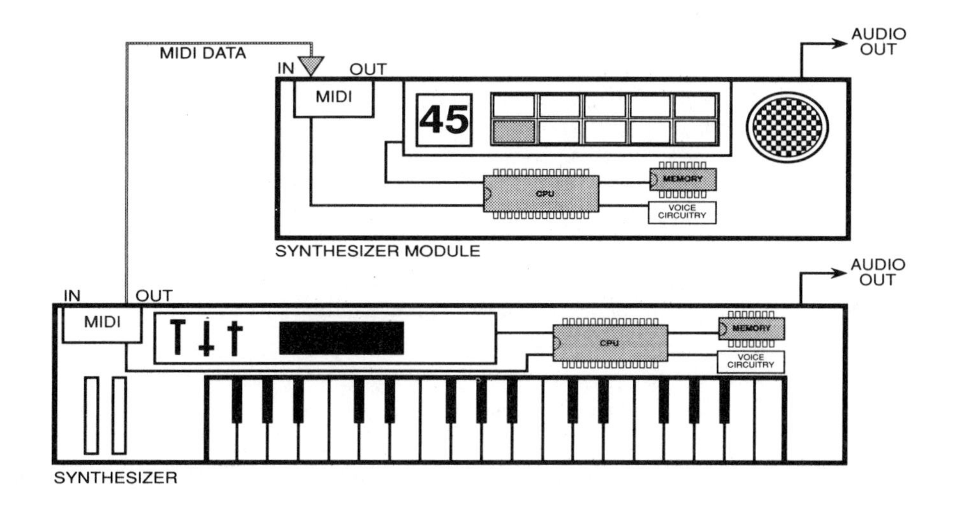

22757 Fig. 4-2

Fig. 4-2. Communication between multiple instruments via MIDI.

For the remainder of this chapter, numerous examples and profiles shall be discussed for each type of the many MIDI-capable electronic musical instruments that are currently available on the market. These MIDI-based instruments are made up of categories which include a wide range of keyboard, percussion, guitar, and woodwind instruments and controlling devices.

Keyboard Instruments

Keyboard instruments are the most commonly encountered electronic musical instruments within MIDI production. This is, in part, due to the fact that these were the first devices to be developed within the industry, and that MIDI was initially developed to record and control many of their internal and performance parameters.

Keyboard-based devices employ a central microprocessor, keyboard, control panel, memory, and general provisions for auxiliary controllers. The two basic keyboard-based instruments are the synthesizer and the digital sampling device.

The Synthesizer

A *synthesizer* is an electronic musical instrument which makes use of multiple sound generators to create complex waveforms that, when combined, synthesize a unique sound character. Each synthesizer tone generator can be controlled with respect to frequency, amplitude, timbre (sound character), and envelope. Modern synths provide digital control over these analog parameters, or more commonly, generate these waveforms directly within the digital domain. Both methods allow the user to store the control parameter values used to generate these sounds within an internal or external memory location as a sound patch.

There is a wide range of synthesizer types on the current market that employ numerous processes for generating complex waveforms. The vast majority of the newer systems make use of complex *digital algorithms (program tables)* to generate their sounds, while the latest generation of devices mix digital synthesis techniques with actual waveform samples to create new and realistic sounds.

The Korg M1 Music Workstation (Fig. 4-3) is an example of a popular system that combines a 61-note digital synthesizer, keyboard controller, and an 8-track internal sequencer into a single system package. This 16-voice device makes use of Korg's proprietary *AI (advanced-integrated) synthesis system*, which can create textures by combining high-resolution sound samples with synthesis technology in the digital domain.

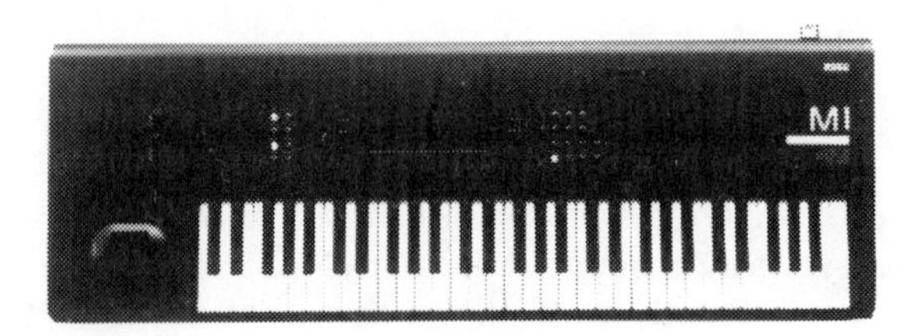

Fig. 4-3. The Korg M1 Music Workstation. (*Courtesy of Korg U.S.A., Inc.*)

The M1 features two megawords (four megabytes), which offer 62 multisampled 16-bit PCM, ROM-based sounds. A complete library of synthesized sounds are factory mapped into 100 preset memory locations, with an additional 100 presets available as user-programmable locations. Additional sounds may be accessed by the system using optional PCM sampled-data cards and ROM cards (containing stored program information). A separate component synthesis system allows users to combine the separate components of pitched or nonpitched sounds to create new voices. *Additive harmonic synthesis* can be used to combine two digital wave generators, which can then be shaped in real time by a high-speed, 16-bit LSI variable digital filter.

These complex synthesized sounds can then be processed using the M1's two independent, programmable stereo digital effects, which may be used in either series or parallel. Up to four of the processors' 33 digital effects can be layered into each program combination and sequence.

In addition to its sound synthesis capabilities, the M1 also features a complete 8-track sequencer which can internally record and play back multitimbral compositions, which can be comprised of synth voices, percussion samples, and digital effects.

The SY77 (Fig. 4-4) is Yamaha's new *flagship music synthesizer* which combines an advanced form of *FM (frequency modulation)* synthesis techniques together with 16-bit PCM sample technology to create what Yamaha calls real-time convolution and modulation. In this way, sampled audio can be used as a basic operator for the internal FM synthesis generators.

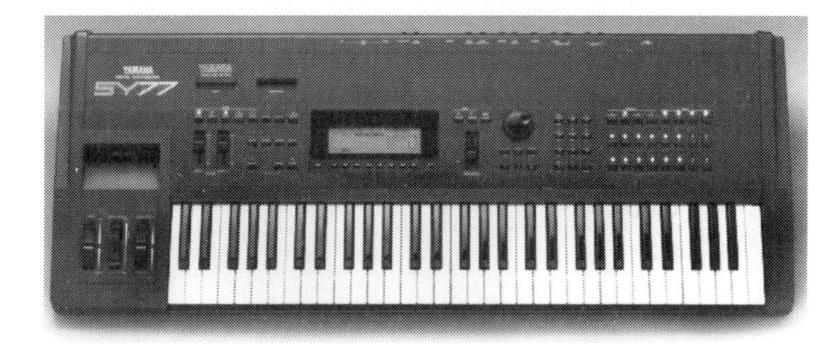

Fig. 4-4. The SY77 music synthesizer. (*Courtesy of Yamaha Corporation of America*)

Technically speaking, this system deals with 16-bit wave data that is sampled at 32 or 48 kHz, using 24-bit internal processing (including digital filtering) and high-resolution analog-to-digital converters. The SY77 incorporates 4 Mb of internal sampled ROM, giving the user a choice of 112 internal waveforms (with an optional external ROM card library for providing additional sounds). Synthesis is accomplished using a 6-operator sound generation system that includes 45 different wave tables, each of which is capable of offering up to 3 independent feedback loops.

This device features 32 dynamically allocated voices (16 for the synthesis section and 16 for the sampling section), digital filtering, dynamic panning, a 61-key velocity, and an after-touch sensitive keyboard. While 64-voice and 16-multipatch presets are offered as user-programmable locations, 128 internal voice presets and 16-multipatch preset locations are used to store factory patches. Also offered are two independent stereo outputs, four independent digital effects processors, a built-in 3.5-inch disk drive, and a 16-track internal sequencer is also included within the SY77.

The Synthesizer Rack Module

Synthesizer systems are also commonly designed as a 19-inch rack-mountable or table-top module system. These devices, which are commonly known as synth modules or expanders, often contain all of the features of a basic synthesizer, with the exception of a keyboard controller. Thus, these modules are controlled through the use of an external keyboard controller or other system (Fig. 4-5).

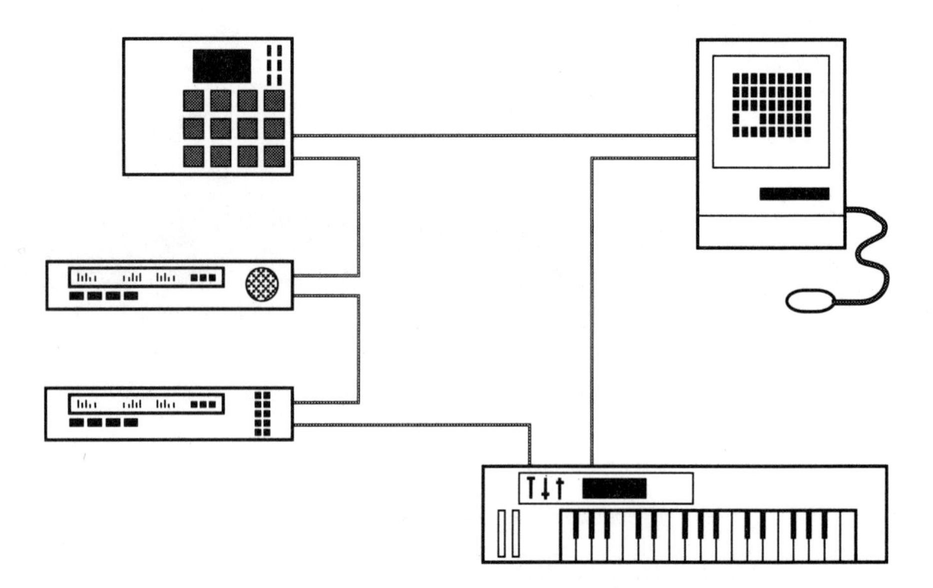

Fig. 4-5. MIDI system configuration that incorporates a synthesizer module.

One such module system is the Korg M3R multiple sound-source module (Fig. 4-6). This 8-voice device is capable of creating and editing internal sounds from up to four separate types of sound sources: ***multisampled PCM data***, ***synthesized-waveform data***, ***extracted-waveform data*** (allowing the extraction or editing of attack transients and nonpitched-noise elements from existing sounds), and ***PCM drum sounds***. These four waveform types have been sampled from various instruments and sound generators and are stored as sound data in the M3R's PCM waveform memory. Additional waveform data can be loaded into the system through two ROM PCM data-card slots, with one being dedicated for program data and the other for PCM data. In addition to supporting its own ROM card library, it is also capable of reading M1 PCM data cards.

Fig. 4-6. The Korg M3R AI synthesis module. (*Courtesy of Korg U.S.A., Inc.*)

The M3R has a program memory capacity of 100 separate sound programs and is capable of storing up to 100 different sound program combinations. It can also be played in one of five possible modes: ***single*** (when a key is pressed, one voice is heard), ***layer*** (pressing a single key results in the output of a number of preprogrammed voices), ***split*** (the keyboard is split so that different voices are heard over different sections of the keyboard's range), ***velocity switch*** (voices change, depending upon how hard the key is struck), and ***multi*** (independent voices being controlled from assigned MIDI channels).

The M3R also includes a built-in independent multi-effects device, which includes 33 different effects types. This allows effects, such as stereo reverb, chorus, delay, exciter, and EQ to be individually layered and cascaded from one effects processor to the other.

The K4R 16-bit digital synthesis module from Kawai (Fig. 4-7) is a 2U rack-mounted system that provides the user with 16 voices within an 8-instrument multitimbral structure. The K4R, like its K4 keyboard controller counterpart, provides 256 internal waveforms. Sixty of these waveforms are PCM samples, while 96 are ***DC (digital cyclic) waveforms*** which are very useful for creating analog-style sounds. Voices can be generated through any of three different types of synthesis: ***additive*** (where waves are mixed together), ***subtractive*** (employing digital filtering), and ***ring modulation*** (amplitude modulation). Up to four different waveforms can be combined at one time and then routed through their own generator stages, which includes a digital low-pass filtering for added signal resonance.

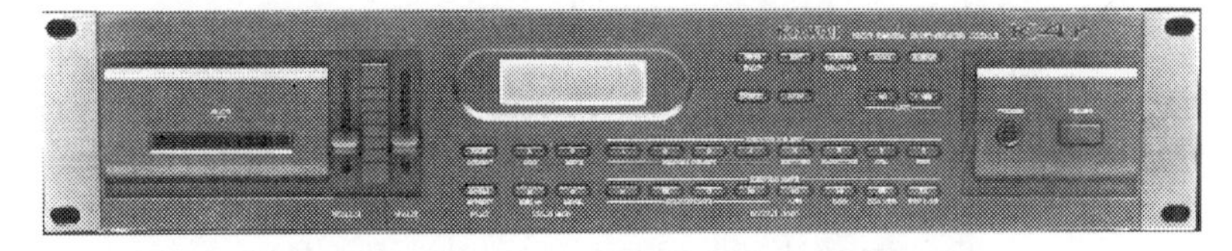

Fig. 4-7. The Kawai K4R digital synthesis module.
(*Courtesy of Kawai America Corporation*)

Combined waveforms can be saved to any of the device's 64 patch-memory presets and up to eight patches can be combined, split, or layered over a keyboard controller, or can be outputted multitimbrally from an external sequencer. Each multipatch setup can be stored into one of the device's 64 available multipatch memories. An optional memory card can be used in conjunction with the device's internal memory for storing an additional 64-patch and 64 multipatch memory locations.

A separate drum-machine voice section is included to provide a wide range of acoustic, electronic, and European-style percussion sounds. Each drum sound is made from the combination of two waves, level, and tunings, and can be individually assigned to any of the keys upon a keyboard. An additional digital effects generator within the K4R contains up to 16 effects settings, including reverb and delay.

Sampling Systems

A *sampler* is a device that is capable of converting an audio signal into a digitized form, storing this digital data within its internal RAM, and reproducing these sounds (often polyphonically) within an audio production or musical environment. Many sampling systems offer extensive edit and signal-processing capabilities, allowing a sample to be modified by the user and saved to a computer disk for archival purposes. These devices also make use of many of the amplification, oscillation, and filtering stages which are found within digital and analog synthesizers. These allow the user to modify a sample's overall waveshape and envelope.

Keyboard-based samplers make use of musical keyboard controllers to trigger and articulate sampled audio by using standard control modifiers, such as after touch and modulation. Once a set of samples has been recorded or recalled from disk memory, each sample within a multiple-voice system can be split across a performance keyboard. In this way, individual sounds can be assigned to a specific key or range of notes. The latter allows a sample or multiple samples to be musically played at various sample rates which correspond to a key's proper musical interval.

Often, newer sampling systems offer a range of signal distribution capabilities, including stereo sampling and multiple outputs (offering isolated channel outputs for added mixing and signal-processing power, or for recording individual samples to a multitrack tape recorder). Many samplers provide data distribution to and from computer-based sample editors through the use of the MIDI sample-dump standard, or by way of a high-speed *SCSI* (*small computer system interface*) *port* for use with high-end sample editors and external hard-disk systems.

The Emax II 16-bit digital sound system from E-mu Systems, Inc. (Fig. 4-8) is an example of a sampler/keyboard combination that offers stereo sampling, synthesis, and sequencing capabilities within a single production package.

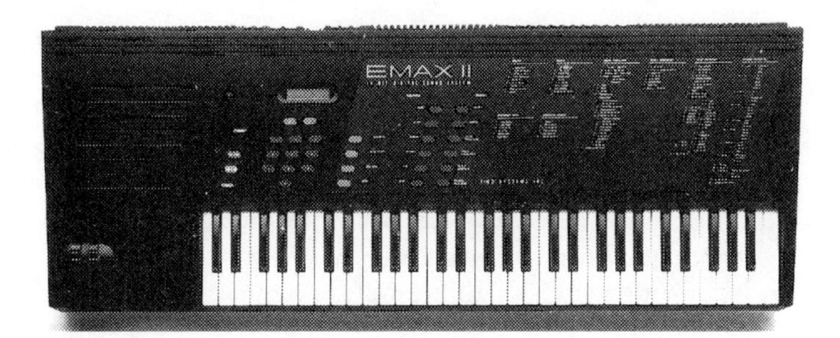

Fig. 4-8. The Emax II 16-bit digital sound system. (*Courtesy of E-mu Systems, Inc.*)

Designed specifically for the reproduction of true stereo samples, the Emax II offers 32 direct audio channel outs which can be configured as 16 stereo voices or 16 monophonic voices with true stereo chorusing. Thirty-two custom digital low-pass filters are employed to provide analog-style timbre control, while 18-bit digital-to-analog converters are used for each of its eight programmable polyphonic outputs. These outputs (configured as four stereo pairs) allow voices to be internally mixed, thereby eliminating the need for a large number of externally dedicated mixer channels to be used within a production setup. Integral sends and returns are also provided for each output, allowing external effects devices to be used without the need for a separate mixer.

Synthesis is carried out through the incorporation of *Spectrum Interpolation Digital Synthesis*. This feature is used to provide a straightforward implementation of additive synthesis techniques to provide sound timbres. System-wide digital signal processing features include:

Basic edit capabilities

Looping

Gain change

Reverse

Digital sample-rate conversion

Digital pitch conversion

Transform multiplication (a function for digitally mixing together two samplefiles)

A 16-track sequencer is also designed into the system for use as a scratch pad or for directly importing compositions from an external sequencer (eliminating the need to carry a separate computer or dedicated sequencer).

The Emax II is available in a variety of configurations, including a rack module and keyboard controller version. Internal memory can be expanded from its standard 1-Mb version to a maximum capacity of 8 Mb, while an optional 40-Mb hard disk can also be added to the system's standard 3.5-inch floppy-disk drive.

An SCSI port offers additional storage and access capabilities. It allows Macintosh-compatible devices, such as optical WORM (write once, read many) and rewritable drives, external hard-disk drives, and computer-based digital waveform editing systems to be incorporated. In order to best take advantage of these features, the Emax II includes a disk operating system that allows access to multiple drives and provides a backup and restore function with user-configurable automated backup routines.

Another such sampling system is the ADS-K 16-voice, 16-bit stereo keyboard sampler from Dynacord (Fig. 4-9). The operating structure of this device unites stereo sampling, disk-drive memory, programmable mixing (with six auxiliary sends and returns), master keyboard features, and digital signal processing within a single unit.

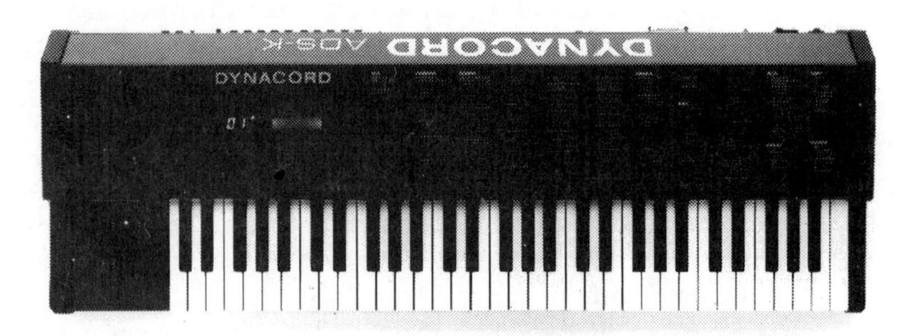

Fig. 4-9. The ADS-K 16-voice, 16-bit stereo keyboard sampler system. (*Courtesy of Dynacord Electronics Inc.*)

The basic ADS version includes 2 Mb of RAM and may be expanded to 8 Mb. Also included is an SCSI port for connecting to an external hard drive or computer system, 2-times oversampling capabilities, 20-bit internal processing, and 16 independent channels. In addition to its included sound library, the ADS is capable of reading all S-900 sample disks, and fully supports the MIDI sample-dump standard. Through the use of a sound fusion feature, up to eight different samples and edited parameters can be superimposed and mixed to create a new sample. An integrated, programmable 8-in/8-out mixer includes such controls as volume, pan, effects send/send return (with pre/post fader), octave select, transposition, and detune capabilities.

The Sampler Rack Module

As with synthesizers, sampling systems commonly integrate all of the necessary signal processing, programming, and digital control structures into a 19-inch rack-mountable unit. Such multiple-voice samplers do not have their own keyboard, but are controlled by the use of an external controller (such as an external keyboard controller, MIDI drum pads, or sequencer).

A popular example of a rack-mounted sampler is the Akai S1000 MIDI stereo digital sampler (Fig. 4-10). This 16-voice sampler is an upgrade of the well-known Akai S900 and features a 16-bit linear sampling format at selectable sample rates of 44.1 kHz or 22.05 kHz. Its 2 Mb of internal memory yields a maximum sample time of 23.7 seconds (mono) and 11.88 seconds (stereo) at 44.1 kHz. Up to three additional 2-Mb memory boards can be added to the system, yielding a total memory capacity of 8 Mb.

Fig. 4-10. Akai S1000 MIDI stereo digital sampler. (*Courtesy of Akai Professional*)

The S1000 can internally store up to 200 different samples, which can be reproduced as a single sampled voice or as a program. A *program* is a combination of existing samples (known as *keygroups*) which can be layered and edited into an overall sound texture. Each keygroup can consist of up to four samples, any of which can be triggered at different velocity levels and defined across any note range on the keyboard.

This device is capable of performing edit and signal processing capabilities, including the programming of up to eight different loop points within a sample, a *join function* (allowing two samples to be merged into a single, new sample), and extensive digital filtering. In addition, a *time-stretch function* makes it possible for samples to be compressed or expanded in time without changing the pitch of the original sample.

The S1000 provides the user with eight individually assignable outputs, stereo outputs, one auxiliary effects send/return, and MIDI in, out, and thru ports. The device's internal 3.5-inch floppy disk is capable of reading and writing either double density or 1.5-Mb high-density disks, and will load sample files that have been recorded using the S900 or S950 samplers. Additional memory options include an internal 40-Mb hard-disk option (S1000HD), an SCSI interface board (enabling up to eight SCSI devices to be interfaced with the device), and an AES/EBU digital audio interface (allowing transfers to and from CD or DAT to be made in the digital domain).

The Roland S-770 is a 24-voice polyphonic, fully digital 16-bit stereo sampling system. This device makes use of 24-bit internal signal processing and 20-bit digital-to-analog converters which supports sampling rates of 48 kHz, 44.1 kHz,

24 kHz, and 22.05 kHz. Extensive digital filtering techniques are employed within this system, which is also capable of resampling filtered sounds while in the digital domain, without having to undergo additional D/A conversion. The S-770's analog output section offers six individual mono outputs, plus a stereo polyphonic output pair. Digital ports provide inputs and outputs for directly communicating digital data with a CD player or *DAT* (*digital audio tape*) recorder.

This device includes a large 64 x 240-dot LCD menu screen, which can be connected to an optional monitor and controlled via Roland's optional RC-100 remote controller or mouse. Memory management includes 2 Mb of internal RAM that can be expanded to a total of 18 Mb. A 2 Mb/1 Mb 3.5-inch floppy-disk drive is included which is capable of reading all current S-series sound libraries for the S-550 sampler. A built-in 40-Mb hard-disk drive allows waveform data to be rapidly saved or loaded (up to 2 Mb of data can be transferred in 3.5 seconds). An SCSI communications port is also included for direct communication with an external Macintosh computer system, hard-disk drive, CD-ROM, or optical-memory storage system.

The ROM Sample Module

Read-only, or ROM, sample technology makes use of prerecorded digital samples that are permanently stored within the device's internal memory chips for reproduction and signal processing. Such sample-based systems differ from a sampler, in that the internal sounds have been recorded and edited at the factory and offer no provision for recording new sounds into memory. Often these devices offer extensive control over waveform envelope and filter editing by the user.

One of the most popular advances in the field of sample technology is the *ROM sample module*, which offers the user a large library of factory prerecorded sound samples. Once selected, a wide range of parameters for controlling these samples (such as envelope, pitch, volume, pan, and modulation) can be modified by the user.

These powerful modules can be used to provide the performer with a large database of actual samples that can be played over the keyboard's range in a polyphonic fashion, often being capable of reproducing up to 32 simultaneous voices.

One of the most notable of the ROM sample modules is the Proteus 16-bit multitimbral digital sound module from E-mu Systems, Inc. (Fig. 4-11). This 2U rack-space device makes use of custom *VLSI* (*very large scale integration*) technology to place some of the most notable sounds from their E-III sound library into a 4 Mb ROM chip (internally expandable to 8 Mb).

Fig. 4-11. The Proteus 16-bit multitimbral digital sound module.
(Courtesy of E-mu Systems, Inc.)

The Proteus offers 32 simultaneous sampled voices that include pianos, organs, strings, horns, guitars, basses, drums, latin percussion, etc. In order to take advantage of layering capabilities, up to eight sounds can be layered onto each key, while the entire device is capable of responding multitimbrally to all 16 MIDI channels simultaneously.

This device features E-mu's MidiPatch, which acts much like a digital patchbay, giving the performer real-time access to over 40 user-programmable parameters that can be stored within any of the device's 192 preset locations (the Proteus XR version offers 384 presets). These parameters are accessible from a MIDI keyboard, MIDI controller, or from the Proteus' internal operating structure. Other features include six polyphonic outputs (configurable to three stereo submixes with fully programmable stereo panning), integral sends/returns, and extensive MIDI implementation.

A Proteus/2 version has also been released which includes 8 Mb of new 16-bit samples, such as solo violin, viola, cello, ensemble strings (both arco and pizzicato), a full range of orchestral woodwinds and brass, harp, celesta, timpani, tubular bells, and a wide selection of orchestral bells. In addition, the Proteus/2 contains a new selection of digital waveform tables for creating a new range of user-programmable sounds.

The MIDI Keyboard Controller

The *MIDI keyboard controller* is a keyboard device that is expressly designed to transmit performance-related MIDI events throughout a modular MIDI system. It contains no internal tone generators or sound producing elements. Instead, it is often fitted with a quality weighted performance keyboard and a control and memory architecture for handling complex MIDI system setups.

Devices of this type make use of internal microprocessors, panel controls (for control over real-time MIDI data and system setup configurations), a full range of MIDI performance, setup, control features, and in certain cases, memory for storing these configurations to internal RAM or disk-based memory.

Generally, keyboard controllers are fitted with a high-quality keyboard and real-time controllers. Often the keyboard will offer an extended key range (66 or 88 keys) which is weighted to resemble the action of a piano keyboard. The keys may also be split into user-defined zones (Fig. 4-12), which can often occupy an area ranging from the full keyboard playing area to a single key. These zones often can be programmed to overlap each other to occupy the same area or cover separate sections of the keyboard. Each zone may be assigned its own MIDI channel, program-change number, continuous-controller functions (often assignable to a modulation wheel, pitch bend wheel, or data fader), after touch, velocity curves, and key transposition settings. In this way, multiple instruments, voicings, and MIDI parameters can be mapped by the user over the playing surface.

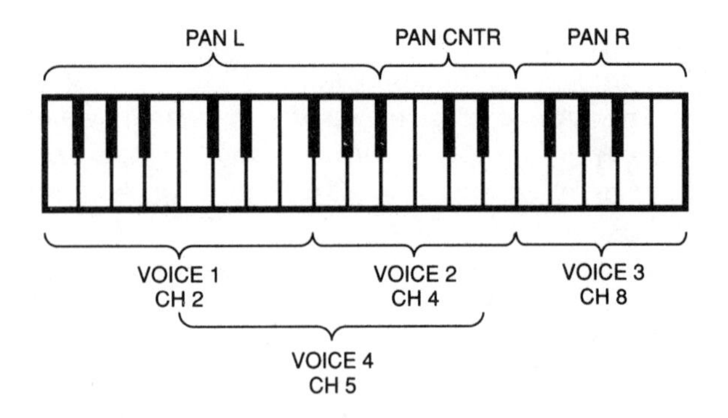

Fig. 4-12. Example of a keyboard that is split into various zones.

The Roland A-80 MIDI keyboard controller (Fig. 4-13) features an 88-key playing surface fully weighted and dampened by a special rotary oil-damper construction. It also features velocity and after touch sensitivity, in addition to the capability for being split into four assignable zones. Each of these zones can be assigned to occupy any individual or overlapping key range and can be assigned its own MIDI channel, program-change number, or continuous-controller message parameters. Zone mute and solo functions operate in much the same way as those found on mixers and recording consoles. The *zone mute* allows a zone to be silenced in real time, while the *zone solo* will allow only the selected zone(s) to be heard at one time. An *extras menu* allows each of the four zones to be given individual setup definitions, such as zone program change, zone main-volume level, modulation, bender information, etc.

The A-80 is equipped with a pitch controller, modulation controller and pitch bender lever, 4 front-panel continuous-controller data sliders, 4-foot controller jacks, and 4 data-control switches. The control sliders can be assigned to any MIDI control message from 0 to 127 and the switches set from 1 to 128. Each controller is capable of simultaneously varying controller messages for each of the four split zones (e.g., slider 1 can control the main volume level in zone

Fig. 4-13. The Roland A-80 MIDI keyboard controller.
(Courtesy of Roland Corporation US)

1 and pan positioning in zone 2, etc.). Any combination of foot pedal (either continuous-data change or on/off switching) can be connected to any of the four foot-pedal jacks.

The A-80 can store up to 64 patch setups within its internal memory, with each patch containing a complete systems configuration, including zone mapping, controller sensitivity and setup configurations, output channel assignments, etc. These patches can then be given names of up to 16 characters in length and placed within a patch catalog. A chain function is also included, allowing up to 32 patches to be linked together in any order for quick access to setups during a performance. An external RAM card is also available, permitting the backing up or storage of additional setup patches.

In addition to controller and keyboard-setup data, the A-80 is capable of storing SysEx data within its patch memory. This allows external sound modules or other controllers to be reconfigured without the need for an external patch librarian.

The MX76 master keyboard controller from Akai Professional features a 76-note (E0-G6) weighted keyboard, and offers programmable dynamic-velocity control, pressure control, and the ability to split the keyboard into as many as four different key groups. Each key group is capable of transmitting its own MIDI event data (channel, program change, velocity, modulation, etc.), in addition to such programmable parameters as key range, key split, transpose, pitch bend, etc.

In order to easily implement these MIDI and control changes, the MX76 makes use of the first 18 keys of the keyboard as *parameter call keys*, which enable the performer to edit any control parameter by simply pressing the appropriate key.

Up to 50 overall-parameter setups can be stored within the system as a library that provides an overall system snapshot of MIDI channels, program changes, key ranges, etc., for each of the four possible key ranges. These library banks may also be arranged into packets that allow up to 20 banks to be accessed in any order during a performance, using the keyboard's up/down keys or from a footswitch. These packets may also be identified using personal notes on the 40-character x 5-line LCD screen.

The MX76 supplies two independent MIDI out ports, over which any of the various key groupings can be assigned. In addition, a MIDI in connector is supplied, allowing data from a sequencer, synth, or other MIDI device to be merged with the controller's data. MIDI clock is also supported allowing tempo and start/stop, as well as continue messages to be sent to external devices.

The MIDI Grand Piano from Yamaha (Fig. 4-14) is a unique keyboard controller that incorporates full MIDI implementation into Yamaha's Conservatory Grand Piano. The MIDI Grand system makes use of fiber optic sensors to accurately capture keyboard expression, without interfering with the normal touch response of the grand piano action. In addition, after-touch sensors, which are mounted below the keys, respond to pressure exerted on the keys after they have been depressed.

Fig. 4-14. The MIDI Grand Piano. (*Courtesy of Yamaha Corporation of America*)

The MIDI Grand's loud (*sustain*) and soft (*una corda*) pedals can also be used as controllers to sustain or soften the sound of external MIDI instruments. Built-in effects controllers include a pitch bend wheel and modulation (which can be controlled by either the system's modulation wheel or by a foot controller). Volume changes can also be handled by the controller's volume slider or by a foot controller.

The controller utilizes two fully independent MIDI processors, each of which are capable of storing up to 11 functional parameters into its 32-location memory bank. A programmable key-limit function enables different voices to be split over the keyboard range. The controller's function mode also allows transposition settings, damper-pedal on/off messages, channel-number assignments, and control settings for all MIDI controllers. A MIDI in jack is supplied, allowing external data (from a sequencer or external controller) to be merged with the controller's data path. Each MIDI processor is also equipped with two independent MIDI out ports for 32-channel support.

The DMP 18 (Fig. 4-15) is a dynamic MIDI footpedal board that offers 18 velocity-sensitive notes (ranging from C to F), which can be assigned to one of seven possible octaves. Upon powerup, the DMP 18 will automatically default to MIDI channel 1, octave 0, key-velocity on, hold off, and poly off. Changes to these defaults can be made by pressing the *program access/play push switch*, which enters the device from the performance mode (*play*) into the programming mode. This allows the performer to change performance parameters in real time by selecting the appropriate MIDI function and by pressing the pedal which refers to the desired parameter value.

Fig. 4-15. The ELKA Professional DMP 18 dynamic MIDI pedalboard. (*Courtesy of Music Industries Corp.*)

Percussion

One of the first applications in sample technology was to record drum and percussion sounds, making it possible for electronic musicians (most notably keyboard players) to add samples of actual drum-sound sounds to their own compositions. Out of this has sprung a major class of sample and synthesis technology that enables artists to make use of drum and percussion voices through the use of a synthesizer, drum machines, or sampler. The latter allows artists to create their own sampled drum and percussion sounds (ranging from traditional instruments to the imaginatively unexpected).

MIDI has brought sampled percussion within the production capabilities of almost every electronic musician, whose performance skills range from basic rhythmic programming capabilities to the professional percussionist/programmer that uses his/her skills for building complex drum patterns.

The Drum Machine

The *drum machine* is most commonly a sample-based digital audio device which is generally not capable of recording audio into its internal memory. Instead it makes use of the playback capabilities of ROM to reproduce high-quality prerecorded drum sounds. Such a device is shipped with factory-loaded ROMs that include carefully recorded and edited samples of the individual instruments that make up the modern drum and percussion set. In addition to these sounds, samples (such as reverberated samples, gated samples, orchestral hits, and James Brown screams) may be added to the list of traditional sounds.

These prerecorded samples can be assigned to a series of individual button pads which are located at the top of the machine, providing a straightforward controller surface that often sports velocity and after-touch dynamic capabilities. Once selected, these drum voices can be edited using control parameters such as tuning, level, output assignment, panning position, etc.

Selected drum sounds may then be arranged into a rhythmic sequence (known as a *drum pattern*). Such drum patterns are generally composed of one or more measures that contain a number of events which are used to trigger drum or percussion sounds according to a specific user- or factory-programmed rhythm pattern. These patterns may consist of a number of basic variations on a playing style, or may represent a pattern that can be taken from an existing library of the many available playing styles (such as rock, country, jazz, etc.).

Most drum machines offer multiple outputs which enable individual or groups of voices to be routed to a specific output. This feature allows these isolated voices to be individually processed at a mixer or recording console (through equalization, effects, etc.) or to record isolated voices onto separate tracks of a multitrack tape recorder.

A basic internal sequencer is also included within most drum machines for the creation and storage of drum patterns. The design and operation of these basic editor/sequencers will often vary widely from one drum machine to the next. For example, certain types require that a defined pattern be performed into the sequencer in real time, while others make provisions for entering note and voice values in step time (Fig. 4-16).

Many drum machines provide a chain function that enables patterns to be linked together into a continuous song. Once a song is assembled, it can be played back using an internal MIDI clock source, or may be synchronously driven from another device (such as a sequencer) using an external MIDI clock source.

In the majority of cases, the individual voices of a drum machine are triggered from a MIDI sequencer, allowing user to take full advantage of the sequencer's real-time performance and editing capabilities. In this way, sequenced patterns can easily be created in step time (often graphically, when used with a computer-based sequencer) and linked together into a song. It is also possible to edit patterns or to merge step- and real-time tracks together to create a more human-sounding composite rhythm track.

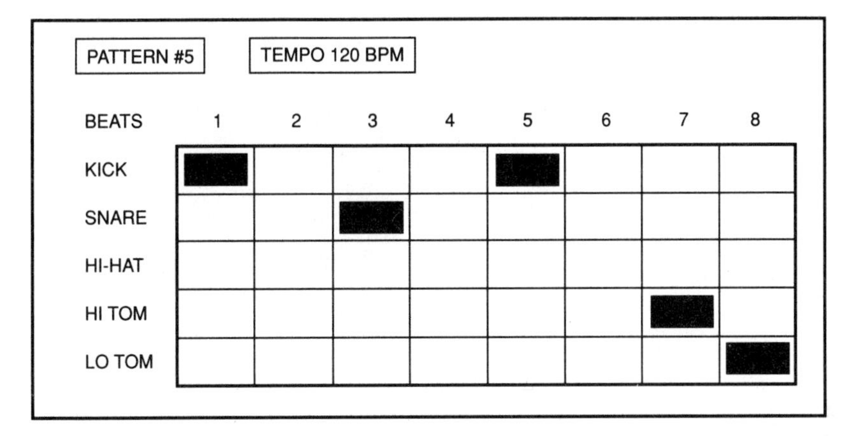

Fig. 4-16. Graphic representation of a drum machine's step-time sequencing feature.

One example of a popular drum machine is the HR-16 high sample-rate digital drum machine from Alesis. This system uses an 18-bit internal processor to provide 49 high-quality 16-bit drum samples, which are hand picked from a collection of drums spanning 60 years and range from pure acoustic tonalities to aggressive power drums and electronic drum sounds. A full complement of cymbal and percussion sounds has also been included.

The HR-16 gives the user 16-digital sample channels, all of which can be individually varied in level, tuned, panned, and assigned to any of four outputs (configured as two stereo pairs). Any of the 49 available voices can be assigned to any of the 16 top-mounted velocity-sensitive drum pads (in eight velocity steps). All 16 voicing/setup parameters and drum patterns can be stored within any of the HR-16's 100 memory locations, and can be chained together to create up to 100 songs or are accessible via MIDI program change. Each pattern can be from 1 to 682 beats in length, and up to 255 patterns can be chained together to create an overall song.

A companion drum machine to the HR-16 is the HR-16B from Alesis (Fig. 4-17). This system is essentially the same in structure to the HR-16, with the exception that it contains 47 sounds that are more aggressive in nature than those offered by the HR-16 and are often composites of several modern-edge sounds. In addition, the HR-16B contains an updated programming chip that functionally operates in controlled tandem with the HR-16, when connected by a standard MIDI cable.

Fig. 4-17. The HR-16B high sample-rate digital drum machine. *(Courtesy of Alesis Studio Electronics)*

Another example of a popular drum machine is the R-8 Human Rhythm Composer drum machine from Roland (Fig. 4-18). This device contains 68 built-in 16-bit sounds (at a 44.1-kHz sampling rate), which includes a basic drum set, Latin percussion, and assorted sound effects. An additional 26 instruments can be created and stored into its internal memory register by modifying existing sound parameters, along with 26 additional sounds that are available when an optional ROM memory card is inserted into the device.

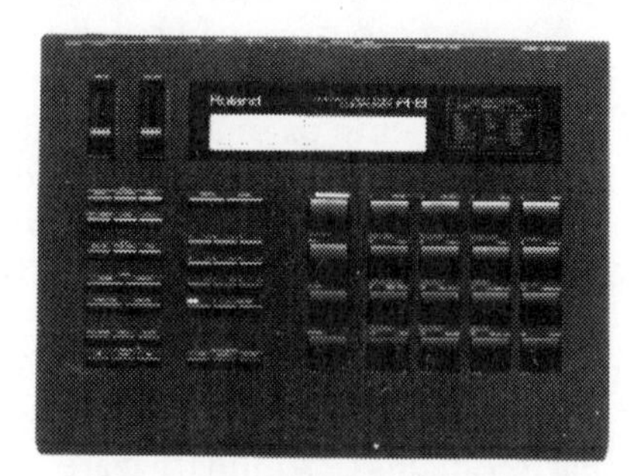

Fig. 4-18. The R-8 Human Rhythm Composer drum machine. *(Courtesy of Roland Corporation US)*

Each of the R-8's PCM sounds can be tuned over an 8-octave (± 4) range and offers control over velocity and decay, while a nuance control enables the user to adjust subtle timbral variations over a 16-increment range for certain instruments. When the nuance feature is assigned to a pad, subtle variations of how each instrument is reproduced is used to add a natural human feel to the overall sound. Any of these four variables can be controlled and programmed in real time through the use of an expression pedal.

The R-8 includes 16-velocity and after-touch sensitive pads, which can be used for inputting rhythm patterns or real-time performance. When using its roll function, control over after-touch sensitivity allows the user to continuously vary the level of a snare drum or cymbal roll.

Sounds may be assigned to any of the pads, and the system makes provisions for saving up to five pad-bank configurations, allowing the user to design and recall custom setups. Each sound on each pad can be assigned its own pitch, pan, and other parameters, while an additional multiassign feature allows one sound to be assigned to all 16 pads, with each being a different increment or user-defined performance variable.

The R-8 incorporates an internal pattern editor that can be used to assemble a pattern with a quarter-note timing resolution of 96 PPQN. These patterns may also be copied, edited, and merged, allowing a new composite pattern to be created. This pattern can then be included within a pattern library and named for later recall. Patterns may then be linked into a song, or a repeat-mark function can be used to loop a predetermined number of patterns to be chained together within a song.

The human-feel function allows minute or substantial variations in velocity, pitch, nuance, and decay parameters to be edited within a pattern or song. In addition to controlling the degree of depth over the course of a song, it can also be randomly used to control a sound or range of sounds. Once created, up to eight feel patches can be stored within the device's internal memory.

Alternative Percussion Voices

In addition to the multitude of sounds that are found within drum machines, percussion voices can be obtained from other sources. Samplers are a common source for obtaining percussion sounds. This is true, both for sounds that are commonly supplied by the factory, or as sample files that are created by the user.

Percussion sounds are also commonly available within the standard ROM presets of newer keyboard and synthesizer modules. These sounds can exist as either PCM (sampled) sounds, synthesized waveforms, or as a combination of the two. Often, synthesizers are equipped with internal signal processors for adding a degree of realism to these voices, without the need for an external processor.

MIDI Drum Controllers

MIDI drum controllers are used to translate the voicings and expressiveness of percussion into MIDI data, where they can be recorded to a sequencer or performed within a live setting. Since the drum sounds within a setup can be directly accessed via MIDI, often a MIDI setup will contain several different methods for controlling and editing a performance or pattern arrangement. The following is a summary of the most popular means of controlling percussion voicings.

Drum-Machine Keypads

One of the most straightforward of all drum controllers is the *drum pad* which is designed into the drum machine. The proper drum-pad voicings and setup parameters are generally accessible by calling up a desired pattern or setup patch from the drum machine's internal memory. Thus, the player need only hit the pads to hear the drum sounds and transmit corresponding MIDI messages. This method provides a certain degree of expressiveness (often communicating velocity and after-touch information).

These controlling pads are generally too small and not durable enough to withstand drum sticks or mallets. For this reason, they must generally be played with the fingers. In addition, their velocity sensitivity is often limited to a range of sensitivity steps and does not express the continuous range of velocity levels.

Keyboard Controller

Since drum machines respond to external MIDI data, controlling devices (such as a MIDI keyboard) can be used to trigger drum-machine voices. There are a few advantages to playing percussion sounds from a keyboard. For example, sounds can be triggered more quickly because the playing surface is designed for fast finger movements and does not require full hand/wrist motions. Another advantage is the ability to express velocity over the entire range of possible values (0–127) instead of the limited velocity steps that are often available with drum pads.

The majority of drum machines allow the user to manually assign drum and percussion voices to a particular MIDI note value. In this way the user can assign notes to a range of drum sounds that fit his or her particular playing style. Additionally, these drum sounds can be assigned to a particular range of notes within a split keyboard arrangement, allowing other sound patches to be simultaneously addressable upon the same keyboard playing surface.

Drum-Pad Controllers

Often within a more advanced production, percussion and drums may need to be expressed with a greater degree of control than can be provided by either drum-machine pads or a keyboard. At such times a dedicated *drum-pad controller* can be used. It offers the performer a larger and more expressive playing surface than can be struck with either the fingers and hands, or with the full expressiveness that can be provided by percussion mallets and drum sticks.

Drum-pad controllers are also capable of being user-programmed, allowing the performer to create various MIDI-related parameter setups. This enables each of the pads to be assigned to a full range of MIDI parameters (such as MIDI channels, velocity levels, one or more note numbers, after touch, etc.), as well as for switching between these program setups within a performance.

The Roland PAD-80 Octapad II (Fig. 4-19) is a controller that provides the user with eight large velocity- and after-touch sensitive pads with additional inputs for six external trigger pads. This device offers features such as layering, pitch bend/modulation control, panning, etc. MIDI implementation allows the performer to assign pad parameters (such as MIDI channel, note number, program change, etc.) to any of its 64 available setup-patch locations (expandable to 128 with the optional M-128 memory card). A patch-chain function is also included within the system, enabling setup configurations to be chained together (along with program-change messages) using any of its eight internal chain tracks. Patch chains can be switched or cycled through using the PAD-80's front-panel selectors or by using an external foot switch.

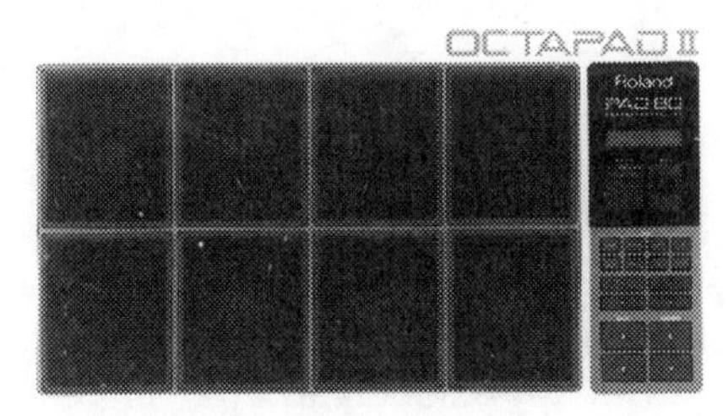

Fig. 4-19. The Roland PAD-80 Octapad II drum/percussion controller. (*Courtesy of Roland Corporation US*)

A layering feature gives the PAD-80 the ability to provide dynamic control over multiple sound textures, and to transmit program changes from any pad in real time. Each pad is able to control up to three note numbers, with each having its own trigger-velocity levels. When in the MIX mode, three voices can be triggered simultaneously, allowing chord progressions to be played from the eight-pad surface. A V-MIX mode allows each voice to be set at a different velocity level, allowing the sound texture to be gradually mixed as the pad is hit harder. Pitch bend and modulation can also be controlled through played dynamics, or through the use of an external pedal. Gate time (the duration that a voice is to sound when triggered) can be set at up to four seconds on each pad.

The drumKAT from KAT, Inc. (Fig. 4-20) is another such velocity-sensitive MIDI drum controller for the professional drummer and percussionist. It consists of 10 natural gum-rubber playing pads, in addition to 9 external trigger inputs. Each external trigger input can be trained to accept a wide variety of input sources. The drumKAT is capable of being adjusted (either manually or automatically) to recognize the dynamics contour of an acoustic trigger, electronic trigger pad, and electronic foot trigger.

Fig. 4-20. The drumKAT MIDI drum controller.
(Courtesy of KAT, Inc.)

The drumKAT's parameters can be configured into individual 32 *KITs* (which specifies the setup function of all selected pads and triggers). Upon selection, it transmits up to six MIDI program changes and main-volume messages. Each KIT can be individually named and up to eight KITs may be chained together into a song.

Each pad and trigger can be programmed using one of three modes: simple, complex, and control.

- In the *simple mode*, each pad and trigger is assigned to a single MIDI voice, and outputs according to the velocity at which the trigger is hit. The note's gate time can be programmed from .025 to 6.3 seconds.
- In the *complex mode*, each pad and trigger may be configured using six additional playing modes.
    - *Note shift mode*: Dynamics control the pitch of a MIDI voice.
    - *Gate shift mode*: Dynamics control the sustained length of a voice.
    - *Velocity shift mode*: Dynamics control which of the three (or all) possible voices will be sounded.
    - *Multiple mode*: Three independent voices will play simultaneously with individually programmable delay times (.025–6.3 seconds).
    - *Alternating mode*: Three independent voices will play one at a time on an alternating basis.
    - *Hi-hat mode*: Three independent voices (two on the pad) are controlled by footswitch position, including a separate sound for the footswitch itself.

- In the *control mode*, MIDI control functions can be directly controlled from the pad surface. These include: program changes, sequence start/stop, pitch bend, after touch, breath control, modulation, sustain on/off, etc.

The drumKAT supports up to 32 MIDI channels over a total of 4 MIDI out ports (2 independent ports for each of the left and right halves of the playing surface), and includes 2 MIDI in ports for external performance control or data-merging functions. In addition, extensive MIDI mapping features are included and the device fully supports MIDI clock with sequencer start, stop and continue, tempo control (including real-time tap-tempo features), and independent click.

MIDI Percussion Controllers

Unlike drum-machine playing pads or keyboard controllers, which are used by most electronic musicians, the *MIDI percussion controller* is generally used by professional percussionists. These systems are commonly designed with a playing surface that resembles a vibraphone. Such a device can be fully configured using its internal setup memory to provide for user-defined program changes, playing-surface splits, velocity, after touch, modulation, etc.

The KAT MIDI percussion controller from KAT, Inc. (Fig. 4-21) is a velocity-sensitive modular system that can be expanded from one to four octaves to suit the player's needs. At the heart of this system is the *master octave*, a device that includes the programming brains, 13-note bars (C to C), 2 function-control bars, and a 2-line backlit display. Up to three additional *expander octaves* may be added to this basic module to create a larger performance range.

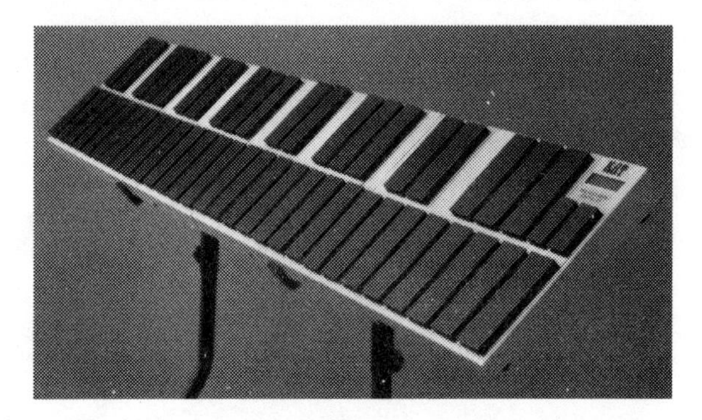

Fig. 4-21. The KAT MIDI percussion controller. (*Courtesy of KAT, Inc.*)

The KAT is able to store up to 256 setups which can be grouped into 32 songs. Each song can contain up to eight setup programs which can be stepped through using a footswitch. Each setup can specify settings for three separate controllers and can transmit program-change and main-volume messages. The first two controllers (which are identical in structure) are called the A and B controllers. They permit the user to program such parameters as note limits, MIDI channel, minimum velocity, maximum velocity, velocity curve (including cross-fading curves), main volume level, transpose, and hold time (.025–6.3 seconds). The third controller (known as the reassignment controller) allows the performer to manually assign MIDI channel and note data to each bar pad on the controller. MIDI and performance parameters can also be controlled from an additional electronic footswitch, in addition to allowing direct control over real-time MIDI commands by striking the appropriate bar pads.

The MIDI Guitar

Guitar players often work at stretching the vocabulary of their instruments beyond the traditional norm, using distortion, phasing, echo, feedback, etc. This has been augmented by developments in pickup and microprocessor technology which have made it possible for the notes and minute inflections of guitar strings to be accurately translated into MIDI data. With this innovation, many of the capabilities that MIDI has to offer (including synthesis, sampling, percussion, and sequencing) are now available to the electric guitarist. Thus, with the implementation of MIDI, a guitar performance can offer traditional guitar sounds, coupled with the advantages of synthesis and MIDI effects technology.

One example of a MIDI controlled guitar/synthesizer system is the Korg Z3 guitar synthesizer and ZD3 guitar synthesizer driver (Fig. 4-22). The Z3 guitar synthesizer makes use of a 16-bit custom LSI and employs two parallel microprocessors for extracting accurate pitch and sound generation of a guitar's strings. This 6-voice device offers 128 custom-preset sound sources with four digital oscillators per voice which can be processed using a digital reverb algorithm.

Fig. 4-22. The Korg Z3 guitar synthesizer and ZD3 guitar synthesizer driver system. *(Courtesy of Korg U.S.A., Inc.)*

The Z3 synthesizer offers two basic operating modes: mode A and mode B. *Mode A* allows the performer to select basic setup parameters (voice select, octave up/down, reverb on/off, etc.). While operating in the MIDI poly mode, this mode allows all 6 strings to control any of the 128 possible voice patches. *Mode B* allows multiple voices to be controlled by individual strings in the MIDI mono mode. When in this mode, patches can be configured in any order with user-assignable parameters.

Foot-controller and pedal-switch options provide control over remote functions (such as real-time control over individual string-parameter settings). The ZD3 synthesizer driver provides control over master volume, program down/up, timbre change (switch), individual string sensitivity, and gain peak indication (for setting sensitivity).

MIDI Wind Controllers

MIDI wind controllers differ from keyboard and many drum controllers because they are expressly designed to bring the articulation of a woodwind or brass instrument to a MIDI performance. Instead of the standard controllers, which are available upon a keyboard (i.e., note, velocity, after touch, and modulation), these controllers often provide touch-sensitive keys, glide- and pitch-slider controls, in addition to a sensor for outputting real-time breath control over dynamics.

These controllers are often capable of creating a dynamic feel which is more in keeping with their acoustic counterparts. Even when incorporating advanced dynamic controllers, a MIDI keyboard is simply not capable of communicating many of the dynamic- and pitch-related expressions (such as breath and controlled pitch glide) that a wind controller is capable of. Often such a controller can be used to add a greater degree of expression and humanism to existing synth patches or digital samples since the instrumental phrasing is often controlled by more dynamic variables.

The EVI1000 electric valve instrument and the EWI1000 electric woodwind instrument from Akai Professional (Fig. 4-23) are controllers which have been developed from the *Steiner Phone*, an instrument created by the renowned synthesist, Nyle Steiner.

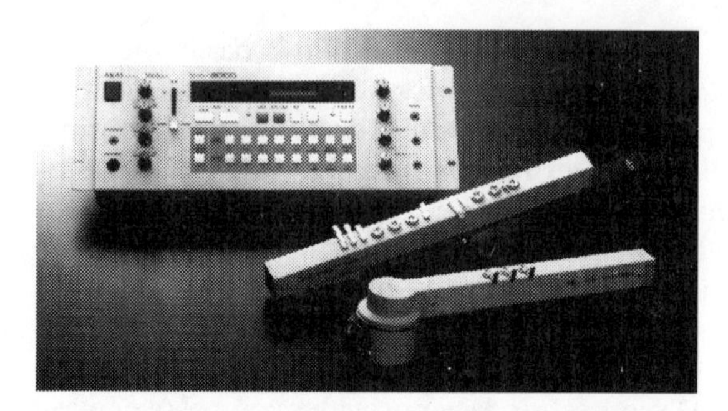

Fig. 4-23. The EVI1000 electric valve instrument and EWI1000 electric woodwind instrument. (*Courtesy of Akai Professional*)

The EVI1000 has six keys (three valve keys and the trill keys) and is fingered like a trumpet, while the EWI1000 has 13 keys and is fingered like a saxophone. Both instruments have an 8 octave range and feature unique touch-sensitive keys that are designed for quick response. A control envelope is derived from an easy-to-blow breath sensor which is mounted within the mouthpiece. The sensitivity of this sensor can be manually adjusted to best match a player's dynamic expression. By biting a lip sensor on the EVI1000's mouthpiece, the amount of glide between notes can be controlled. Glide on the EWI1000 is controlled by a touch-responsive glide sensor. Pressure on the lip sensor of the EWI1000 controls the amount of vibrato and modulation to create effects such as *growl*, *wah*, and *tremelo*. A vibration-control lever on the EVI1000 is used to control these same effects. Both instruments feature a touch sensor that allows the performer to bend notes up or down, in addition to a touch-sensitive octave roller.

The accompanying EWV2000 sound module contains two programmable monophonic analog synthesizer voices. This module includes such programming features as: wave envelope FM, complex filter shaping, modulation and pitch-bending capabilities, 64 programmable patch banks, auto tuning, a 16-character backlit LCD display, and an external sound input for use with an external sampler or synthesizer.

Summary

In conclusion, there is a wide range of MIDI-based musical instruments on the market from which to choose, and each type is able to provide the performer with a specialized function or range-of-sound capabilities.

It is important to keep in mind that although this chapter has presented the reader with examples of currently available models, each of these instrument types and controller combinations are constantly being updated and improved by musical equipment manufacturers. This fact often requires that the musician keep abreast of present advances in music technology to take full advantage of the improved and/or expanded capabilities which these devices can offer.

Chapter 5

Sequencing

Apart from electronic musical instruments, one of the central devices found within MIDI production is the *MIDI sequencer*. A sequencer (Fig. 5-1) is a digitally based device that is used to record, edit, and output performance-related MIDI messages in a sequential fashion.

These MIDI-related channel and system messages commonly represent such real- or nonreal-time performance events as note on/off, velocity, modulation, after touch, program/continuous-controller messages, etc. Once a performance has been recorded by a sequencer (in the form of a sequential stream of digital MIDI messages), these events are then stored within the device's internal memory, where they may be edited and (whenever possible) saved to hard and/or floppy disk for archival purposes. When playing back such a sequence, the device will then output these MIDI event messages in the same order as when they were originally recorded.

Most sequencers have a certain design similar to their distant cousin, the multitrack tape recorder, in that MIDI data can be recorded onto separate tracks which contain isolated and yet related performance material that is synchronous in time. However, unlike its cousin, each of these individual *tracks* can be assigned to any MIDI channel and can store large numbers of performance- and control-related data (within the memory constraints of the sequencer). Upon playing back such a multitrack sequence, MIDI instruments and devices, which are assigned to a specific MIDI channel number, will only respond to any sequenced track that is likewise assigned to this channel number.

The number of individual tracks offered by various sequencers vary widely from one manufacturer and model type to another and range from 8 to in excess of 500 tracks. The majority of current sequencing systems are capable of transmitting data over 16 simultaneous MIDI channels, although only a limited number of professional sequencers are capable of communicating MIDI messages over two independent MIDI data lines, giving them the capability of addressing up to 32 MIDI channels.

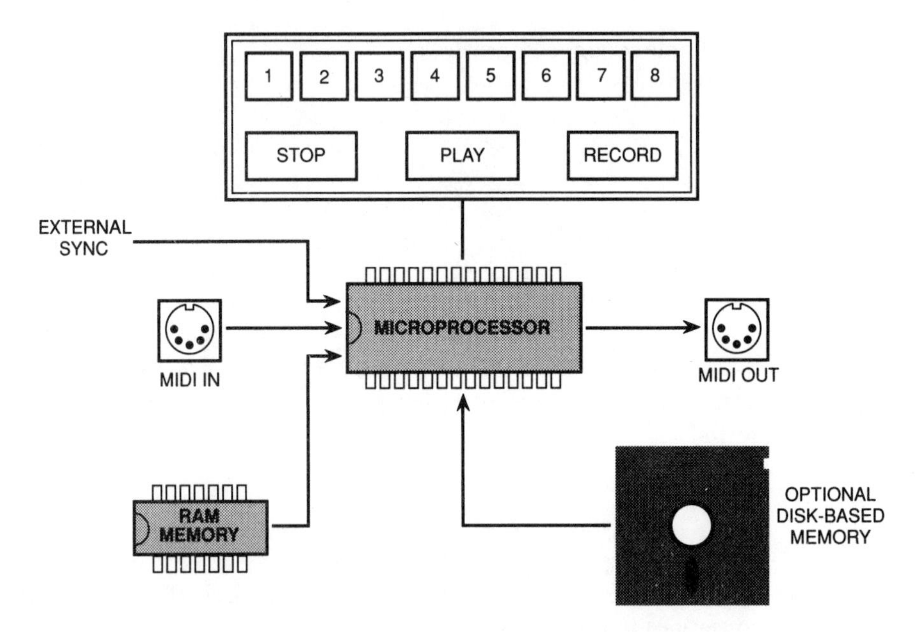

Fig. 5-1. Basic functional diagram of a sequencer.

A Basic Introduction to Sequencers

Unlike the multitrack recorder, where actual audio signals are recorded along the length of a magnetic tape, a sequencer is used to digitally store and manipulate MIDI performance data (and not the musical sound itself) within a block of random access memory (RAM). Like computer-based word processors and most other hardware/software systems, the storage of data and extensive control over this data within computer RAM allows MIDI-encoded musical passages to be easily manipulated using common computer cut-and-paste editing techniques, as well as more complex data processing procedures.

Memory

While the sequencer is not used for storing sounds, but is instead used to instruct MIDI instruments and devices as to what channel, note, velocity, program, etc. is to be played, a great deal less digital memory is required over its audio-recording counterparts (such as is required for digital audio). However, the amount of music, timing, and controller-related data which can be recorded into a sequencer is determined by the size of the device's internal memory. This capacity will often vary from one sequencer to another and, in certain cases, it may be expandable (as with the personal computer).

Some sequencers are limited to a note capacity of less than 3,000 notes, while others offer capacities in excess of 100,000 notes. This might sound like large numbers, but when you consider the amount of notes which can be played within a complex song and multiply that over a number of simultaneous tracks, it is surprising how fast memory can be eaten up.

The amount of memory used within a song is most notably dependant upon a number of MIDI events that are recorded into a sequence. For example, a song which is jam-packed with fast riffs and triplets will require more memory than would a song of equal length that is sparse and meandering. In addition to the simple storage of note-on/off data. The continuous-controller functions (such as pitch bend, after touch, and controller faders) can generate large amounts of data within a short time to communicate accurate real-time controller movements, and thus will use up additional memory space.

Recording

Commonly, a sequencer will be used as a workspace for the creation of personal composition. This often means that a musical passage will be played into the sequencer one track at a time, while playing a MIDI controller. For example, an artist might get a great idea for a song and, using a synth keyboard, lay down a basic track. Once done, a bass keyboard line might be laid down behind the basic riff on another track, then a drum machine and MIDI controller pads could be used to lay down a basic drum beat. This process could continue until the basis of a song begins to develop.

Certain sequencers offer the capability for recording all of the MIDI channels within a multitrack sequence at one time. This feature enables several performers to record a sequence in one live pass, or allows all channels of a completed sequence to be transferred from one sequencer to another in real time.

Almost all sequencers provide facilities for *punching in* and *out* (Fig. 5-2) within a previously recorded sequence track. This function is used to drop in and out of record upon a selected MIDI track in real time, allowing the artist to perform over a flawed musical passage any number of times until the performance is acceptable.

Although punch-in/out points can often be performed manually *on-the-fly*, many sequencers are capable of performing these punch functions automatically. This is often accomplished by entering the measure numbers into the software to mark where the in/out points are to be executed. Once done, the sequence can be rolled back to a point just before the punch-in measure, allowing the artist to play along with the sequence while the sequencer performs all of the necessary in and out switching functions.

In addition to the recording of performance data in real time, it is also possible to enter note values in *step time*. This refers to the ability to enter notes and other events into a sequence one note at a time. Step time is useful for such

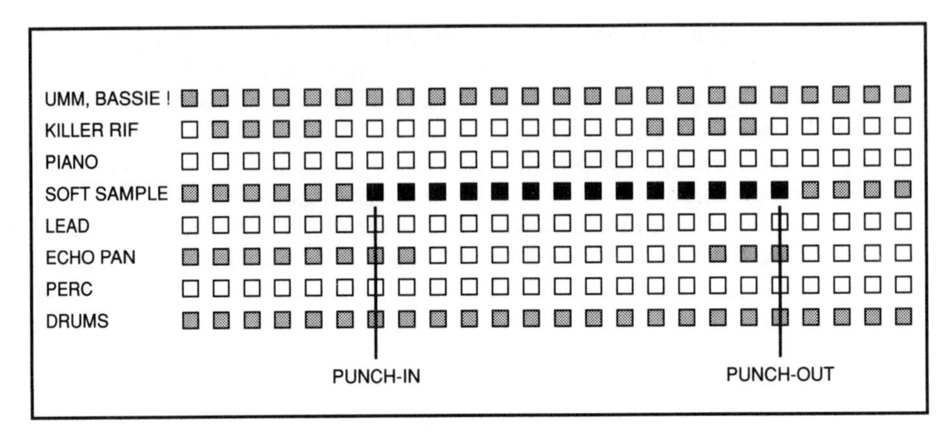

Fig. 5-2. Sequencer screen showing punch in and out points.

purposes as entering more notes into a passage than can be manually played, for entering new notes into an already recorded passage, or for restoring notes that are inadvertently chopped off during the edit or overdub process.

The means by which step time is entered into a sequence will vary from sequencer to sequencer, and will generally allow various parameters (such as note duration, velocity, event type, etc.) to be selected by the user. Such sequencers allow individual notes to be entered into a sequence by first selecting their notational duration value (e.g., quarter note, sixteenth note, etc.). Using this method, once a value is selected, all that is required of the performer is to play each successive note (most commonly upon a keyboard) until he or she wishes to enter a rest value, or new note of a different duration value. This type of step sequencer will also permit the audition of each note in a stepped fashion to allow for editing of any previously entered note or value.

Certain sequencers will also allow segments of sequenced sound to be looped into a sequence in a drum-machine fashion. Such devices enable the user to input and edit a track segment that falls directly within the boundaries of a specified number of measures. Once done, this segment can be programmed to loop, while at the same time be recorded onto another continuous sequence track. This feature allows a basic track (such as a drum pattern) to be laid down, while freeing the player to perform additional fill and/or accent parts in real time.

Playback

Once a sequence is composed and, if possible, saved to disk, all of the individual sequence parts can be distributed to the various MIDI instruments or devices over any of the 16 available MIDI channels. Since MIDI is used to transmit real-time performance messages to the various MIDI instruments within a setup, it is a simple matter to selectively change the actual sound-producing elements of each instrument during the *playback* of a sequence. This powerful patch selection feature gives the artist the ability to alter, experiment with, and re-orchestrate a musical soundscape at any time.

In addition to these advantages, the digital world of sequencing permits the random movement from one measure within a sequence to another without the *fast-wind time* associated with tape. Another advantage is the ability to change the tempo of a sequence without affecting the pitch. In this way, the tempo of a sequencer may be varied at any time without effecting the compositions original key.

Quantization

Within a sequence it is possible to experience timing errors from two possible sources: internal digital timing errors and human-performance errors.

A typical sequencer uses the quarter note as a reference for breaking time down into discrete clock intervals. These devices will then further divide the intervals into a resolution of 24 clock *pulses per quarter note* (*PPQN*). This resolution will break a measure into 96 equal divisions. However, for a rhythmically complex and demanding composition, many sequencers are capable of higher resolutions, such as 48, 96, 192, and 480 PPQN.

By far, most common timing errors begin with the performer. Fortunately, "to err is human," and standard performance timing errors often give a piece a live and natural feel. However, for those times when timing goes beyond the bounds of nature, it is possible to make use of the sequencer's digital capabilities of correcting timing errors through the use of *quantization*.

Quantization allows timing inaccuracies to be adjusted internally within the program to a nearest desired musical time division (such as a quarter, eighth, or sixteenth note). For example, when performing a passage which requires all involved notes to fall exactly on the quarter-note beat, it's often easy to make timing mistakes (even on a good day). Once the track has been recorded using quantization, it is possible for the sequencer's microprocessor to recalculate each note's start and stop times to reproduce precisely within the closest time division (Fig. 5-3). Such quantization resolutions often range from full whole-note to sixty-fourth-note values.

Upon quantization, certain sequencing systems will actually change the note-on/off times within its memory, making it impossible to reverse the quantization process or requantize to a greater resolution. Other sequencers, however, will retain the note timings that were originally recorded into the sequence, while processing quantization timing upon playback in real time. This method allows the performer to change or remove quantization at any time.

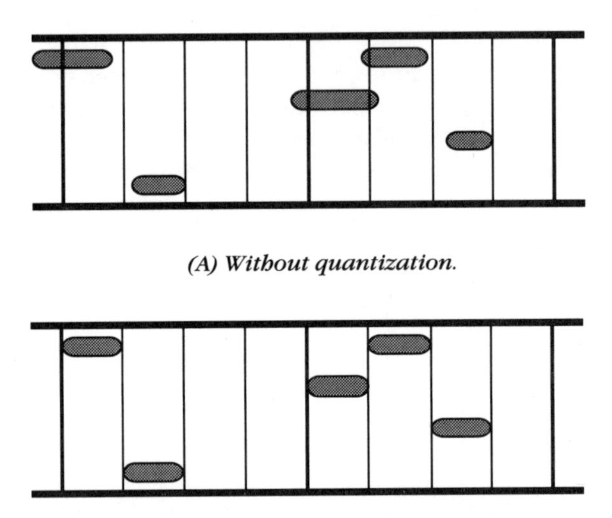

(A) Without quantization.

(B) Quantized to a quarter-note resolution.

Fig. 5-3. Pattern showing quarter-note values.

Caution should be taken with some sequencers when quantizing sequenced tracks that contain continuous-controller data; the sequencer may attempt to bunch large numbers of these events to the nearest interval. This will have the initial effect of rendering this controller data as useless, while possibly causing additional timing problems during playback due to data *bottleneck*.

Programmable Tempo

An added advantage to the reproduction of sequenced data over real-time tape playback is the fact that a sequencer's playback tempo can be easily altered without changing pitch or other real-time parameters. This allows the performer to change the tempo to best match the feel of the song or lower the tempo to play along with previously recorded tracks, etc., more easily.

Some sequencers are capable of being programmed to automatically change tempo within a song (known as *tempo-mapping*). As a MIDI sequencer makes use of MIDI clock timing signals to transmit sync signals to other MIDI devices, a sequence that has been tempo-mapped will be properly reproduced in sync by all of the instruments and devices within a system. However, when syncing a tempo-mapped sequencer to a non-MIDI device (such as a multitrack tape recorder), it will be necessary to make use of a "smart" SMPTE-to-MIDI interface that can be programmed to maintain sync in step with these tempo changes.

Editing

A central feature offered by most sequencers is the ability to edit a sequence in numerous ways. Edit features and capabilities will often vary from one sequencer to another and include the ability to audition and edit a sequenced track in step time to change its pitch, start time, and duration. Most sequencers offer standard cut-and-paste editing techniques, which allow a segment of sequenced data to be cut, copied, or reinserted at any point within a track or onto any other track.

Program or continuous-controller messages can also be inserted and changed in the edit mode. This method allows for control over instrument or device program and parameter changes within a sequence that has previously been recorded.

Jukebox Performance Sequencers

A number of sequencers are available which offer the option of playing whole sequencer files (one after the other) in a jukebox fashion. This application is often used for stage bands or individual artists who make use of MIDI to provide live-backing instrumentation.

This feature permits the user to enter sequences into a songlist, along with a desired time for pausing between songs. Once on stage, the sequencer can automate the entire music set. It can often graphically cue the musician to what song is up next and enable the artist to manually pause or skip between sequences.

Sequencer Systems

Several types of sequencer systems are currently available to the user. Each type offers a unique set of advantages (such as physical integration within musical instruments, portability, power, etc.). It is also true that the different sequencer types have their own basic operating feel, which may be preferred by one electronic musician while another person might have an inclination towards another type of working system. This *feel* will often vary from one sequencer and/or manufacturer to another (even between those that fall within the same working category). When buying a sequencer, it's always a good idea to shop carefully, keeping in mind your personal working habits, present and future growth needs, and the feel or ease of the system.

The Internal Sequencer

Many newer and more expensive keyboard synthesizers and samplers are being designed to offer an *internal sequencer* that is integrated with the instrument's internal microprocessor. Such a sequencer is capable of recording performance data without the need for external MIDI peripherals, and it often offers track selection and such tape-related transport controls as record, play, and stop/pause buttons.

Internal sequencers often have the disadvantage of not offering edit facilities to the user beyond multiple track and punch-in/out record commands. However, since they are designed into the keyboard instrument itself, they can be easily used on-the-road, or when the inspiration bug hits. Quite commonly, these devices are capable of communicating MIDI data to and from other instruments within the system. They are also capable of recording sequences from an external sequencer to eliminate the need for taking an external sequencer on the road. Certain keyboard systems employing an internal sequencer and floppy disk allow sequences (and often patch configurations) to be saved to disk for archival purposes.

The Hardware-Based Sequencer

Hardware-based sequencers are stand-alone devices which are designed for the sole purpose of MIDI sequencing. These systems make use of a dedicated operating system, microprocessing system, and memory that is integrated with controls for performing sequence-specific functions. These sequencers often vary in the number of MIDI tracks that are available to the user. Certain systems provide a limited number of tracks which can be instantly accessed by pushing a dedicated track button, while others employ a numeric keypad for accessing a larger number of tracks.

Hardware-based sequencers commonly emulate the basic function of tape transport (record, play, start/stop, pause, etc.) in addition to fast-wind and locate commands for moving quickly to a specified point within a song. These locate commands are commonly used in conjunction with manual and automatic punch in/out for overdubbing tracks at various points. Many such devices offer more extensive editing features than are available within an internal sequencer, commonly offering both real- and step-time editing, note, velocity, program change, copy and track merging capabilities, tempo changes, etc.

A hardware-based sequencer commonly displays programming, track, and edit information through the use of an LCD. This display type is often limited by its size and resolution, and is generally limited to information that relates to one parameter or track at a time.

The MMT-8 multitrack MIDI recorder from Alesis (Fig. 5-4) is an 8-track hardware-based sequencer that is designed to resemble the control environment of a multitrack tape recorder. The 8-track buttons, combined with the play, record, fast forward, and rewind buttons, allow immediate recognition as standard recording controls.

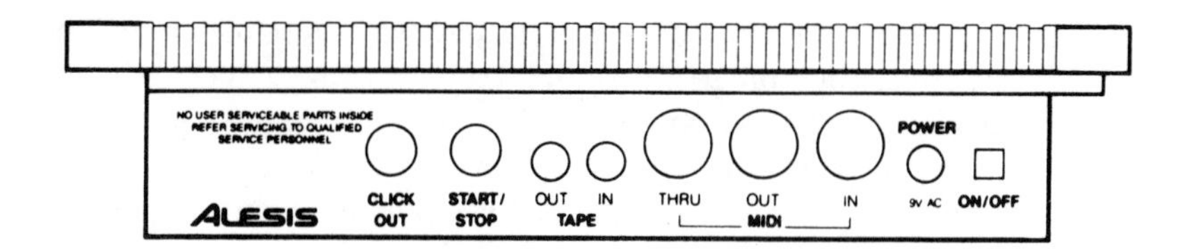

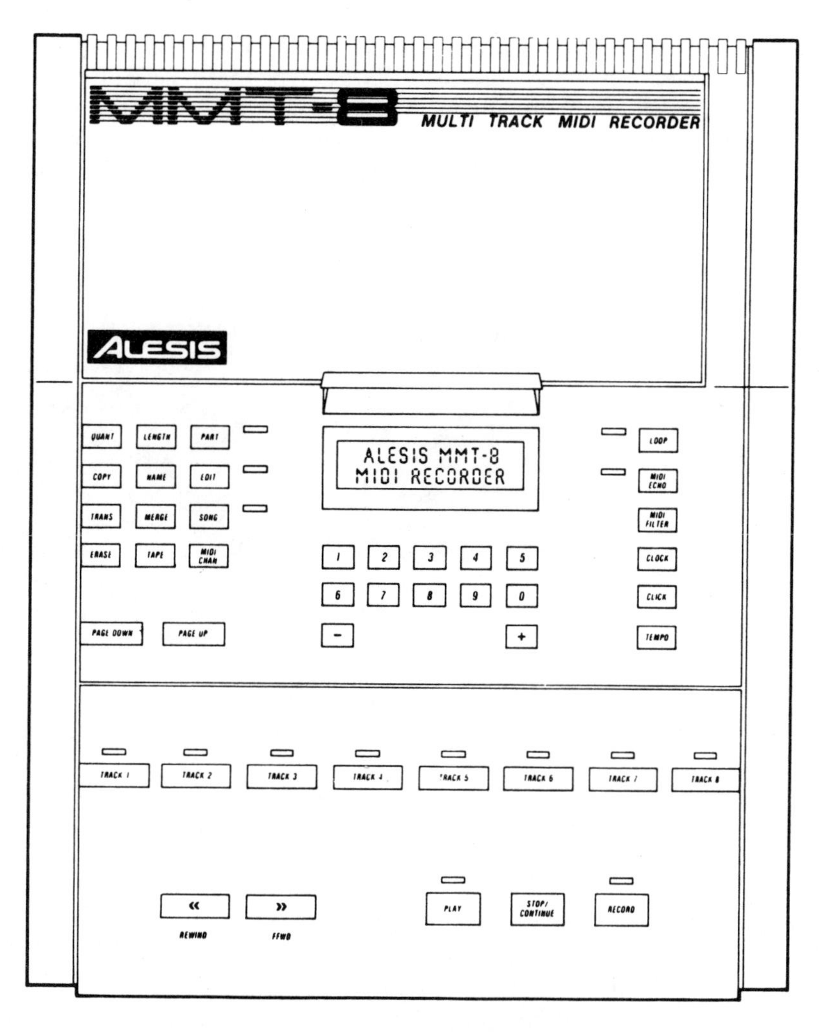

Fig. 5-4. Alesis MMT-8 multitrack MIDI recorder.
(*Courtesy of Alesis Studio Electronics*)

The MMT-8 has a memory capacity of approximately 9,000 to 11,000 notes which is available to any of the 8 tracks, with each track being responsive to any of the 16 available channels. This device arranges individual 8-track sequences into a memory space, known as a *part*. There can be up to 100 parts within its internal memory (0–99), and these may be chained together into an arranged group of parts (known as a *song*). In addition, up to 100 songs may be placed within the MMT-8's internal memory. For example, with these chaining features, the user can record a sequenced part, one verse (part), chorus, bridge, etc. at a time, and then assemble these parts into a final edited song version.

The MMT-8 also includes a full range of editing capabilities, such as note transposition, quantization, copy, and erase features. This device can be directly synchronized to tape, using its rear tape in/out jacks. However, it is also capable of responding to SPP messages through the use of the MIDI in port.

The Kawai Q-80 is another hardware-based sequencer (Fig. 5-5) that offers up to 26,000 notes of memory and up to 320 tracks. In addition, it includes a disk drive for storing data. Unlike most other sequencers, this device incorporates static RAM memory in its design, meaning that all sequencer data is retained in memory, even when the system's power is off.

Fig. 5-5. The Kawai Q-80 digital MIDI sequencer.
(*Courtesy of Kawai Musical Instrument Mfg. Co. Ltd.*)

The Q-80 is capable of storing up to 10 songs into memory, any of which can be saved or loaded to disk. These songs can also be chained together into an extended musical performance. Each song may contain a total of up to 32 tracks, while up to 16 of these tracks may be recorded upon at once. Tracks may be recorded in either real or step time, and a tempo track is also available for adjusting tempo during a song.

The Computer-Based Sequencer

Sequencers are also available as software packages (Fig. 5-6) which make use of the personal computer for performing central processing, memory, and I/O (input/output) functions. These systems are capable of making use of the extensive versatility that a computer offers in the way of speed, digital signal processing capabilities, memory management, and its ability for performing a diverse range of tasks under software control.

Fig. 5-6. Computer, interface, and computer-based sequencer package. (*Courtesy of Voyetra Technologies*)

Sequencing software is available for most Apple, IBM/compatible, Yamaha, Atari, and Commodore personal computers. The majority of these programs require the use of an external MIDI interface, which enables the computer to send and receive MIDI data (with the exception of the Yamaha C1, Atari ST, and Megas, which have a built-in interface).

There are several advantages to computer-based sequencers over their dedicated counterparts. Among these are their increased graphic capabilities, allowing visual and direct control from a keyboard or mouse over a multitude of tracks. In addition, they use graphics as an aid in editing tracks using standard computer cut-and-paste techniques. Graphic-pattern editing also allows the user to easily change the pitch, start, and duration times of a note, as it appears on the screen, often through the simple movement of a mouse.

Computers are digital devices that are designed for memory management. This type of sequencer is designed to easily store files onto either hard or floppy disks. The note capacity is restricted only by the amount of available internal RAM memory and this is often user expandable.

Computer-based sequencers generally incorporate a graphic-screen environment for inputting MIDI data in real time, by displaying an active MIDI track window or screen. This visual interface allows the user to instantly access the active track (or tracks) that are to be recorded upon. By using the sequencer's transport control, the correct start measure can be easily located and the sequencer placed into the record mode (Fig. 5-7). Such graphic environments for accessing and recording onto sequencer tracks will often vary between computer system and program manufacturer.

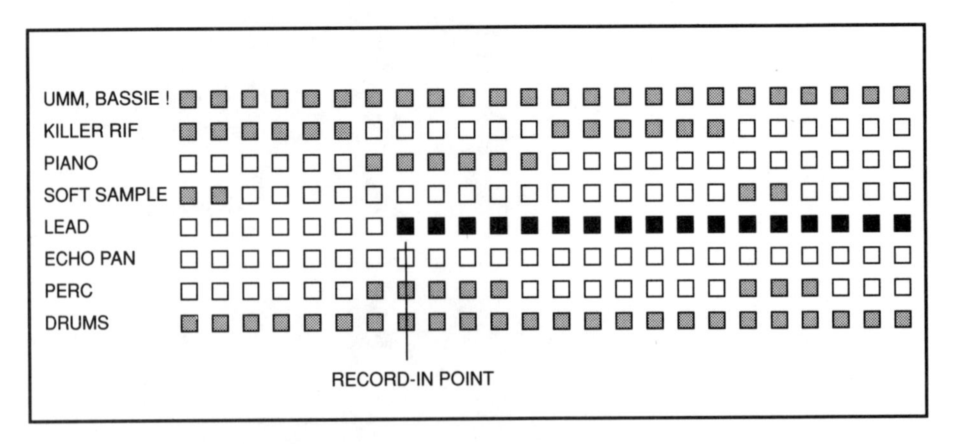

Fig. 5-7. An IBM-style example for recording onto track 5 starting from measure #8.

Performance data can also be entered into these programs in nonreal time, through the use of a piano-roll or drum-pattern type of edit window. Such a window allows the performer to input note-related data directly onto a grid which directly displays note values (vertical) over time (horizontal) in a continuous piano-roll fashion. Depending upon the software, note, start time, and duration may be input into the system from a MIDI controller (musical keyboard, etc.), computer keyboard, or through the use of a mouse.

Editing

One of the strongest aspects of the computer-based sequencer is the advanced editing ability that commonly results from extensive digital signal processing and graphic capabilities. Using standard random access cut-and-paste methods, a musical segment can easily be moved from one track to another, or a musical passage can be cut from a song and saved to a clipboard (within internal RAM) for later use. A passage within a track can also be copied (effectively creating a repetitive loop).

Using the pattern or piano-roll style edit screens, a range of notes can be edited as they appear on the computer screen using either the keyboard or mouse. A popular means of changing the pitch of an existing note is to place the cursor over the current note value (Fig. 5-8A) and manually raising or lowering the pitch. This can be done either from the keyboard or by clicking a mouse key and manually moving the note to its new pitch value. In a similar fashion, it is also easy to change a note's start time (Fig. 5-8B) or alter the duration of a note (Fig. 5-8C).

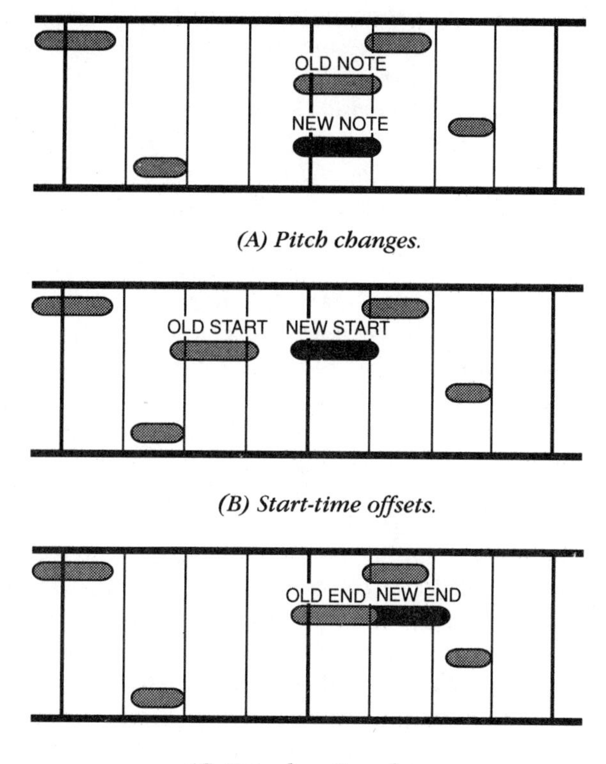

(A) Pitch changes.

(B) Start-time offsets.

(C) Note-duration changes.

Fig. 5-8. Basic note-editing capabilities.

A note's velocity can also be directly changed from the edit screen through the use of a number of program-dependant methods. Certain programs provide a graphic window which show the individual velocity values. These can be changed by simply moving the mouse cursor over a velocity indicator and dragging them to a new value. Other program types may require that velocity parameters be changed outside of the confines of the edit screen.

Often other methods are possible for changing a range of velocity values within a track. As with the previous method, certain sequencers permit the rescaling of a range of velocity values by drawing a curve within the velocity window to graphically represent the desired dynamics. Other systems provide a less interactive means of rescaling by offering a limited number of value changes (such as set, adjust, fade, etc.).

Control over MIDI events (such as program change and continuous-controller messages) are commonly available as a computer-based edit function. Depending upon the computer type, MIDI events can be edited in various ways: either as an edit function embedded within the sequencer package itself, or by accessing an external program which offers direct, graphic control over event type and values (often existing as an onscreen data-fader program).

Facilities for filtering or globally editing basic parameters may also be included within the software, or accessible through the use of an additional program. Such capabilities might include the ability to filter out parameter changes, such as program change, pitch bend, after touch, SysEx, etc.

The capability to merge two or more tracks is a tool which enables a performer to polish and edit individual tracks. Once they have been polished to his or her liking, they can be merged into a single composite track. It should be kept in mind that once a series of tracks has been merged (and the source tracks have been erased), they usually cannot be easily unmerged. As an exception to this, should the notes' range be far enough apart in pitch, certain sequencers will allow a *split* function to be performed, which can again separate the merged notes back to individual tracks.

As previously noted, one of the beneficial aspects to computer-based sequencing is its ability to integrate with other software programs. Within MIDI music production, two program systems that are often indispensible are the patch editor and patch librarian. These programs allow the user to create his/her own sound patches, and arrange and store large volumes of patch-bank datafiles onto hard-disk memory. Patch librarians (which are used to distribute voice patch and setup data to the various instruments within a MIDI system) can be used as a stand-alone program. However a number of sequencer packages are capable of integrating a librarian directly into their software.

In addition to storing patch data, it is also possible for a librarian to distribute patch and setup parameter data to each instrument or device within the system. This arrangement makes it possible for an entire MIDI setup to be configured before playing a music sequence.

As with all sequencer types, computer-based sequencers often vary widely in form and function. They will generally vary with the type of computer the sequencer is designed to interface with. Each package type offers distinct advantages and disadvantages, and the choice is often governed by personal budget, production requirements, and taste. The following example outlines information on several popular computer-based sequencing packages.

Performer Version 3.4 from Mark of the Unicorn offers a graphically-oriented professional sequencer package for the Macintosh family of computers. This compositional/editing program interfaces graphically with the user through the use of moveable windows (Fig. 5-9), which allow quick and easy access to various control features. Each window offers a set of *mini menus* for performing window-specific tasks. The following is a brief explanation of basic window functions.

- *Edit menu*: This offers edit commands, such as snip, splice, merge, and shift (to move events forward or backward in time). A repeat feature can be used to paste multiple copies of a passage into a sequence, and a filter editor allows for MIDI data to be filtered within a track or sequence.

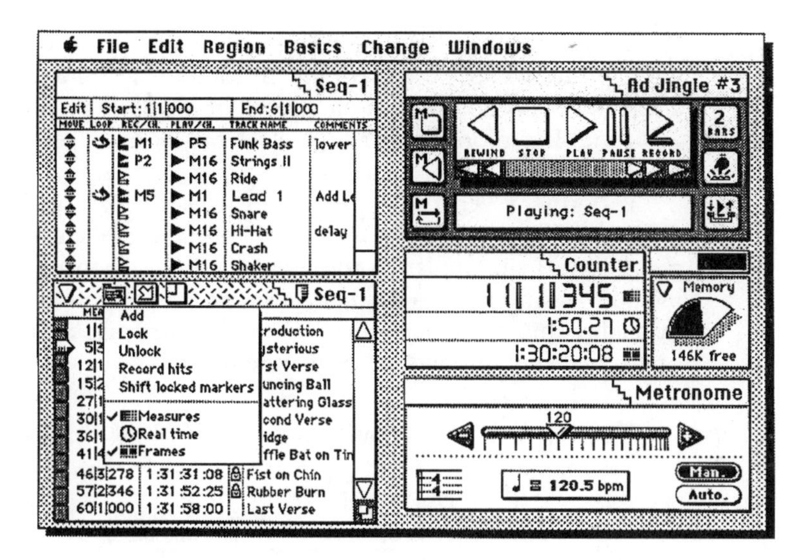

Fig. 5-9. Main screen of Performer sequencing software.
(*Courtesy of Mark of the Unicorn*)

- *Region menu*: This menu allows for a series of commands (such as change velocity and controller data, split notes and time scaling) to be performed upon a range of events from a short passage to multiple sequences.
- *Basics menu*: Features selections for step recording, input patching and filtering, MIDI interface and transmit/receive sync settings, metronomic functions, and MIDI system reset commands.
- *Change menu*: Offers selections for global meter, tempo, key changes, and looping functions.
- *Tracks window*: Allows track names, channels, and comments to be assigned for an unlimited number of tracks up to 32 channels. Multiple tracks may also be recorded into the sequencer over different channels.
- *Markers window*: This window allows MIDI events to be placed into a cue list at precise SMPTE times, measures, beats, or ticks.
- *Metronome window*: Allows tempos to be manually selected or conducted into a tempo map that contains multiple tempos and meters.
- *Counter window*: Displays the current sequence position in any combination of measures, real time, and SMPTE frames.
- *Controls window*: Allows Performer to be manually operated using familiar tape-deck style controls.

Editing can also be done using Performer's graphic-editing screen, which offers piano-roll style event editing. In addition, a continuous data grid appears at the lower portion of the edit screen for displaying pitch-bend data, velocity, and controller information, any of which may be easily reshaped through the use of a mouse.

Performer's Version 3.4 provides the user with a notation-editing window (Fig. 5-10) which allows the user to edit sequenced data using a standard music-notation format. Similar to the graphic-editing feature, the notation window includes a time ruler, a markers strip, and a continuous data grid. However, instead of a pitch ruler and note grid, this window displays note values for editing on a grand staff in standard music notation.

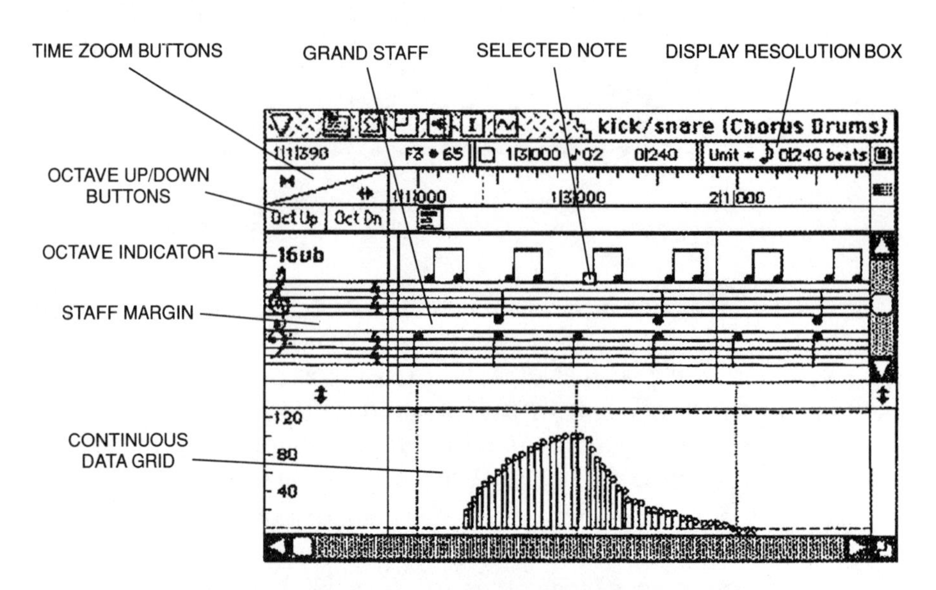

Fig. 5-10. Detail of Performer's notation-editing window. (*Courtesy of Mark of the Unicorn*)

Performer also makes use of a structure for grouping multiple blocks of sequence data (known as *Chunks*) into a larger composition in a simple building-block fashion. Chunks (Fig. 5-11) contain a set of sequenced tracks (such as a verse and bridge) and can be placed into the song window in any order and at any beginning time. These Chunks may be linked together either sequentially or overlayed in time for complex musical expression. This feature has advantages within video and film scoring because any Chunk can be easily slipped in time, thus creating a simplified means of matching the timings of sequenced tracks to picture.

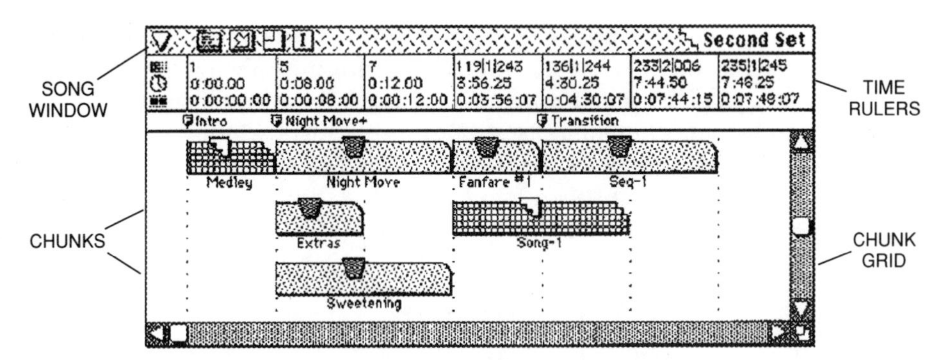

Fig. 5-11. Performer's Chunking feature for chaining sequenced events. *(Courtesy of Mark of the Unicorn)*

Data-controller faders are implemented within Performer for the purpose of providing a graphic tool for manipulating continuous-controller information directly from the Macintosh. Each fader (Fig. 5-12) is capable of generating continuous-controller messages, which can be used for applications, such as MIDI mixing (velocity/panning), adjusting instrument parameters in real time, controlling MIDI-controlled effects devices, etc.

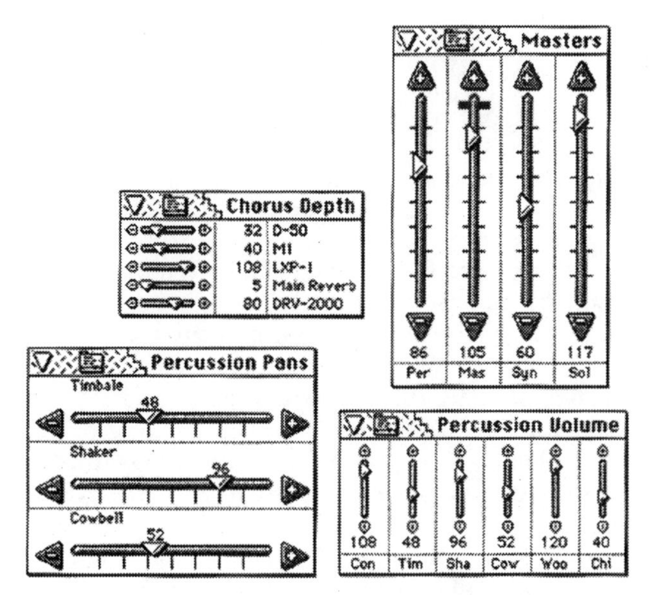

Fig. 5-12. Performer's advanced control slider applications. *(Courtesy of Mark of the Unicorn)*

Vision Update 1.1 (Fig. 5-13) is a professional sequencer package from Opcode Systems, Inc. for the Macintosh family of computers. This system may also be arranged using graphic control windows, such as the control bar, file, sequence, editing, list editing, and faders windows. Each Vision file may be recorded as one continuous sequence or comprised up to 26 sequences. These sequences can then be chained together into a song through the use of an editable sequential list.

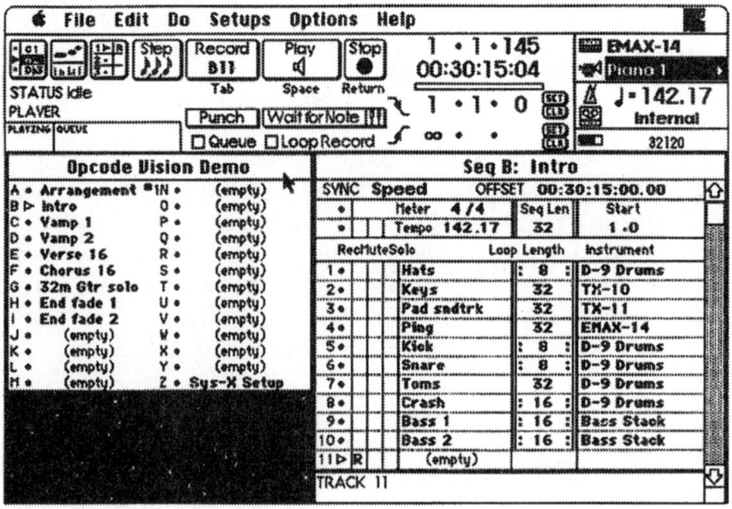

Fig. 5-13. The main screen of Vision 1.1 with the new program change field selected. (*Courtesy of Opcode Systems, Inc.*)

Vision's graphic-editing window (Fig. 5-14) is represented by a standard graphic piano-roll window that makes use of a bouncing ball or scrolling line to show which note is playing. A note's pitch, duration, or velocity can be changed in a single stroke by moving the mouse in predefined ways. A list-editing window enables discrete MIDI events to be edited as a SMPTE time-encoded list environment for use with special controller events or visual sound effects.

Vision is also capable of communicating directly with any Opcode librarian, allowing it to import patch-program names which can be saved with the sequence file. It is also able to *subscribe* to patch files that are created in Galaxy (Opcode's universal editor/librarian). This enables the program names within a sequence to be automatically updated after being reorganized or changed within Galaxy.

Up to 32 moving, animated graphic faders are available for direct control over MIDI continuous-controller data. These faders may be recorded directly within a sequence in real time, or as a snapshot for inserting a scene change at any time.

Sequencer Plus Version 3.0 (Fig. 5-15) offers a series of network tools for the IBM/compatible, which are available in three sequencing packages (SP1, SP2, and SP3). The basic sequencer is incorporated into all three packages. It is comprised

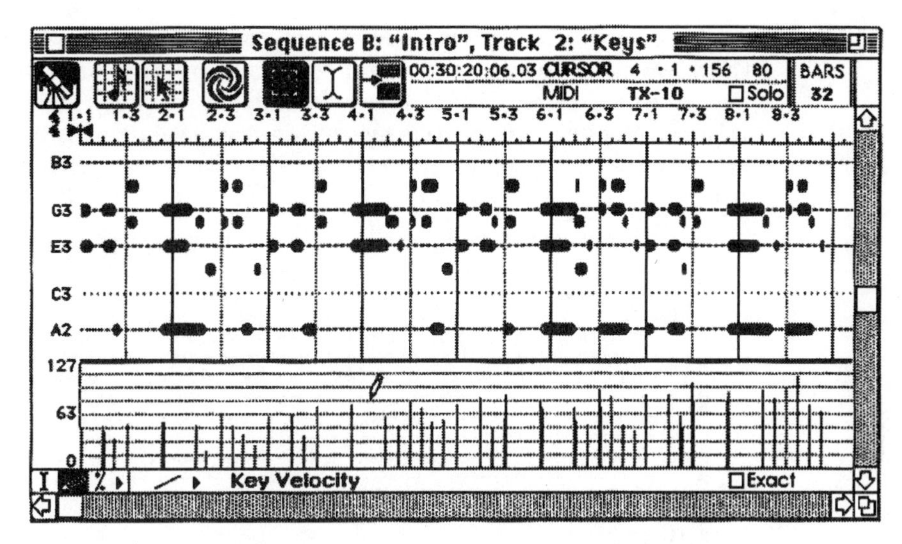

Fig. 5-14. The graphic-editing window, showing the pencil tool in the strip-chart editing velocity. (*Courtesy of Opcode Systems, Inc.*)

of independent polyphonic tracks which incorporate a *Trackscan* feature that allows the user to set the number of tracks to suit production requirements (up to 500 for SP1/2 and over 3000 for SP3). By assigning an instrument per track, a song can be created one part at a time, in either real or step time. Sequencer Plus includes full mouse support and may be externally synchronized to FSK, MIDI sync or clock, and MIDI sync with SPP.

Main

Song TAKE 1 STOP Mem 221248
Tk 3 Electric Piano BPM 114 CK: INTERNAL 26:0

Trk	Name	Chn	Grp	Prg	Trans	Quant	Loop	Mute	Offset		Bars
1	Drums	8	A	49	——	——	——	——	——	1	109
2	Bass	6	A	26	——	——	——	——	——	2	112
3	Electric Piano	1	A	11	——	——	——	——	——	3	112
4	Brass	3	A	2	——	——	——	——	——	4	51
5	Alto Sax	5	A	22	——	——	——	——	——	5	101
6	Tenor Sax	4	A	62	——	——	——	——	——	6	107
7	Trumpet	2	A	12	——	——	——	——	——	7	101
8	Guitar Solo	7	A	54	——	——	——	——	——	8	109
9	Background Brass	3	A	2	——	——	——	——	——	9	102
10	Melody (tenor sax)	4	A	62	——	——	——	——	——	10	102
11	——————	1	A	11	——	——	——	——	——	11	0
12	——————	1	A	11	——	——	——	——	——	12	0
13	——————	1	A	11	——	——	——	——	——	13	0
14	——————	1	A	11	——	——	——	——	——	14	0
15	——————	1	A	11	——	——	——	——	——	15	0
16	——————	1	A	11	——	——	——	——	——	16	0

Main Menu

Record Delete Mute Loop Name Solo Tempo Chase GROUP EDIT FILES
VIEW XLIBRARIAN OPTIONS PUNCH-IN H_MULTI Quit

Fig. 5-15. Sequencer Plus Version 3.0 main screen. (*Courtesy of Voyetra Technologies*)

Songs may be edited on a macro level, using the view screen (Fig. 5-16) that shows which measures in the recorded tracks contain MIDI data. Songs may also be edited on the micro level by zooming into any measure on any track with the use of an edit screen (Fig. 5-17) that allows viewing and precise manipulation of MIDI data.

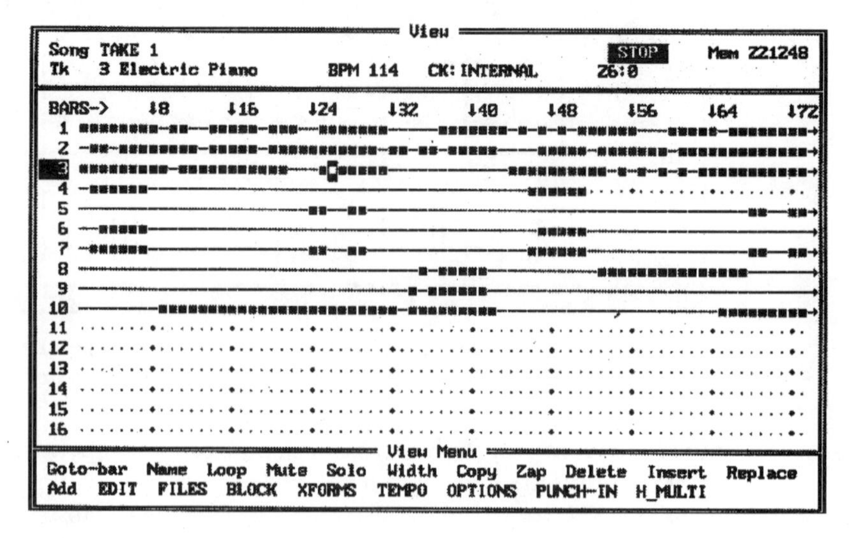

Fig. 5-16. Sequencer Plus Version 3.0 view screen. (*Courtesy of Voyetra Technologies*)

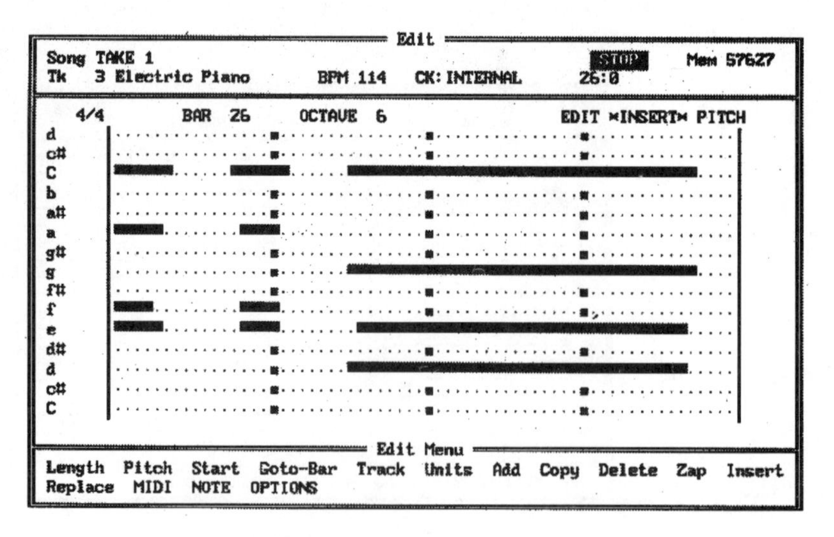

Fig. 5-17. Sequencer Plus Version 3.0 edit screen. (*Courtesy of Voyetra Technologies*)

The Sequencer Plus also provides a series of data transform functions consisting of such note-editing features as transposition, setting/adjusting of note velocities, harmonic inversion (inverts the track over a specified note axis, while keeping them in the same key), map, merge, offset, quantize, etc.

Each of the three packages incorporate a network organizer that is capable of loading up to 32 instruments with banks of patch data from the PC's hard disk/diskette, and saves instrument settings globally with each song. However, SP2 and SP3 incorporate a universal librarian within the package that allows the system to manage and archive over 100 instrument models from over 20 manufacturers. The SP3 incorporates an additional *MIDI data analyzer* which acts as a diagnostic tool, displaying MIDI data either onto a grid screen (showing a busy stream of channelized data), bulk hex screen, or formatted trace screen (which provides a simple English translation next to each command).

Textures Classic from Magnetic Music (Fig. 5-18) is a pattern-based sequencer for the IBM/compatible that allows up to 96 patterns to be grouped into a complete song. Each pattern may be a discrete sequence that is up to 24 tracks deep and up to 2730 beats in length (27.3 minutes long at 100 bpm). Once created, patterns can be linked together (with up to 99 links being available in the song mode).

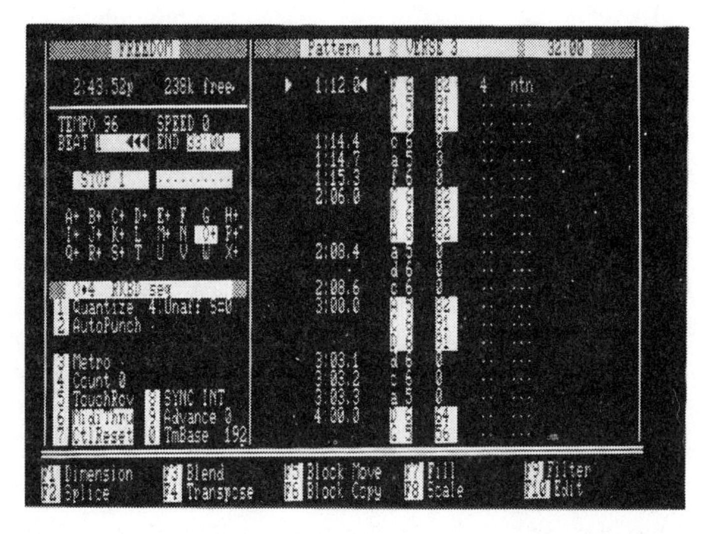

Fig. 5-18. Textures Classic sequencing program. (*Courtesy of Magnetic Music*)

Keystroke and pop-up dialogue windows are provided within this program for access to transport control, files, patterns, tracks, link editing, etc. The edit window provides standard cut-and-paste block editing, with general edit capabilities including, transposition, velocity, continuous-controller messages, fill command (for the creation of repetitive sequences), chord inversions, etc. Sync modes enable the sequencer to lock internally to tape (FSK or SMPTE) or to MIDI clock with SPP.

Standard MIDI Files

As a result of the wide range of computer-based sequencing programs and computer systems currently on the market, a standard has been developed which enables sequence files to be freely interchanged between different programs using the same or different computer systems. This format (known as *standard MIDI files*), encodes time-stamped MIDI event data, as well as song, track, time signature, and tempo information. Standard MIDI files are able to support both single and multichannel sequence data, and are supported either through the load/save or import/export functions of many popular sequencer packages.

In addition to the ability to import and export a sequence to another software/computer type, standard MIDI files have made it possible for shareware sequences to be included on computer bulletin board services which maintain a library of sequenced songfiles. A partial listing of these bulletin boards can be found in Appendix C.

Drum-Pattern Editor/Sequencers

A number of computer-based editors and sequencers, specifically designed for programming drum patterns using a straightforward graphic format, are available. These mouse-driven software programs commonly generate a grid pattern onto the computer screen, which places the instrument voices (MIDI notes representing the various drum sounds) along the vertical axis, while time is represented along the horizontal axis. Individual drum patterns may be built up by clicking on a number of desired drum sounds at specific times within the measure. Once created, this single drum pattern can be repeated, or a varied number of drum patterns may be created and linked together into a song. Should changes in pattern order be desired, the sequence in which the patterns appear to best fit the song can be altered.

Drum-pattern editors commonly offer a wide range of features, such as the ability to change MIDI note values (thereby changing drum voices), adjust note and pattern velocities, save sequence as a MIDI file for export to an external sequencer, and synchronization to an external sequencer.

UpBeat from Intelligent Music (Fig. 5-19) is a unique rhythm sequencer for the Macintosh family of computers. This 32-track MIDI recording and graphic-editing tool provides the user with a rhythm library which can associate drum machine or keyboard sounds, patterns, and setups with descriptive user-defined names, rather than as patch or pattern numbers. UpBeat allows the performer to enter into record at any time within a sequence and (after an initial setup) will automatically record directly to the appropriate user-named track that is dedicated to the specific MIDI instrument being played.

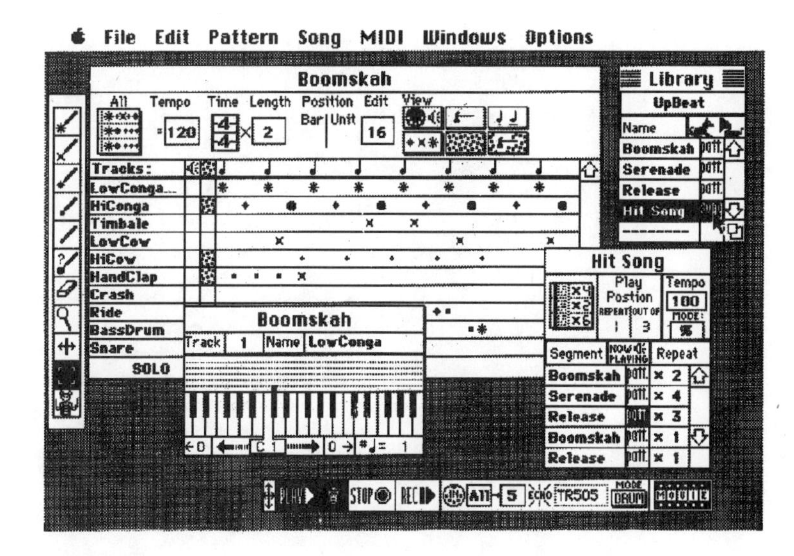

Fig. 5-19. Window for the UpBeat rhythm sequencer. (*Courtesy of Intelligent Music*)

This program provides graphic tools which allow the user to edit and manipulate rhythmic patterns, and to generate random accent variations which add a human feel to the music. UpBeat also allows sections of music to be painted directly to the screen with a mouse. These sections can then be chained in a variety of ways to create or link songs.

Coda's MacDrums is a slightly different type of drum-pattern editor for the Macintosh computer. Unlike most editors, MacDrums (Fig. 5-20) is a self-contained, fully programmable, 4-voice polyphonic drum synthesizer and sequencer. This software package operates by allowing the user to program up to 35 percussion instruments in a standard editor fashion. No device other than a Macintosh is required. The drum sounds are samples of the actual instruments which may be played directly from the computer's speaker or through an external speaker system. This system is also MIDI compatible and features assignable MIDI channels, note values, and velocity. Each of these are editable during playback.

Drummer is a drum-pattern editor from Cool Shoes Software designed to work on an IBM/compatible computer. This straightforward program (Fig. 5-21) offers full control over each instrument name, MIDI channel, and MIDI note number within each pattern. Pitches and overall loudness can also be adjusted within each pattern and individual notes can be adjusted to one of 10 velocity settings. An auto-fill function may also be used to add additional notes to a basic pattern to give it a more human feel. The program's score page allows for up to 2000 patterns to be linked into an arrangement and may be saved as a standard MIDI file for export into a sequencer or notation program.

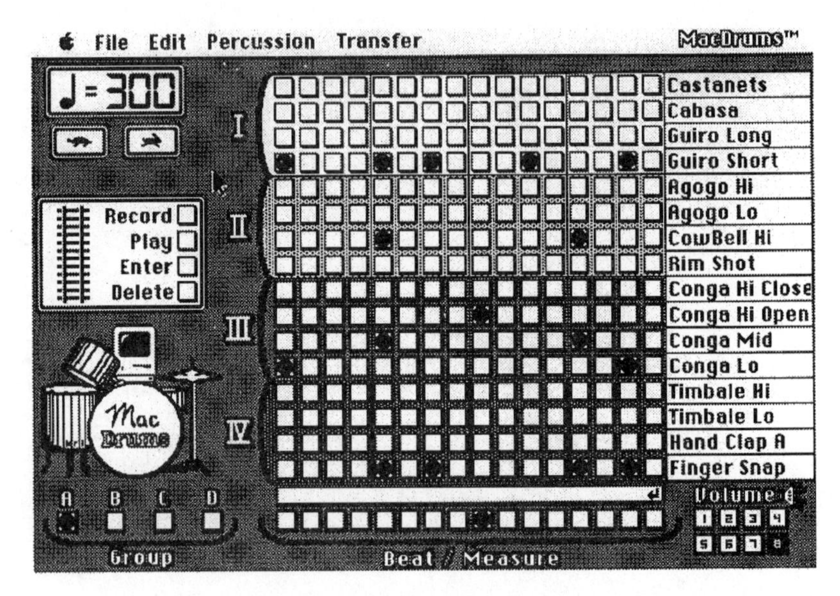

Fig. 5-20. MacDrums drum synthesizer and sequencer program. (*Courtesy of Coda Music Software*)

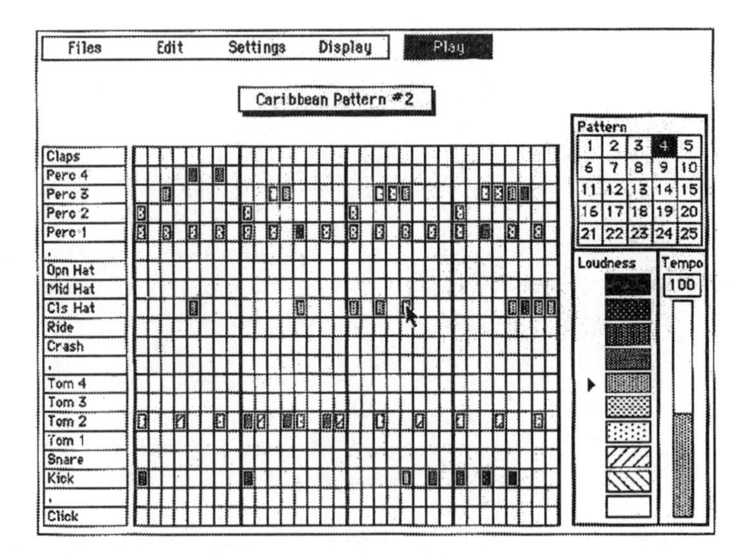

Fig. 5-21. Drummer drum-pattern editor. (*Courtesy of Cool Shoes Software*)

Cue-List Sequencing

Sequencing software is most commonly designed to provide a high degree of control over music-related MIDI note functions (note on/off, note value, velocity, etc.). However, when placing dialogue, sound effects, and music cues within an audio, video, or broadcast production, there are times when the need for controlling a series of discrete or dynamic events will not follow the rules of music sequencing. Instead, a specialized approach to MIDI production is required. Such time-related events are often best placed within a sequence as a series of time-related cues. The graphic representation and control over such cues is often known as *cue-list or list-edit sequencing.*

A cue-list sequencer places data that relates to a specific cue or range of cues (sample trigger, program changes, controller messages, MIDI mixer movements, etc.) within a vertical sequential-edit list. Each line of this list contains data pertaining to time (often SMPTE time) and event-related information.

Certain top-of-the-line sequencers have begun to integrate cue-list editing into their software packages. An example of a dedicated cue list-based sequencer is the Q-Sheet A/V Version 2.0 from Digidesign (Fig. 5-22). Using appropriate MIDI equipment and a Macintosh computer, this SMPTE MIDI-based software package is able to trigger and automate such devices as samplers, synthesizers, MIDI effects devices, MIDI mixers, or any other device which can be controlled by MIDI commands.

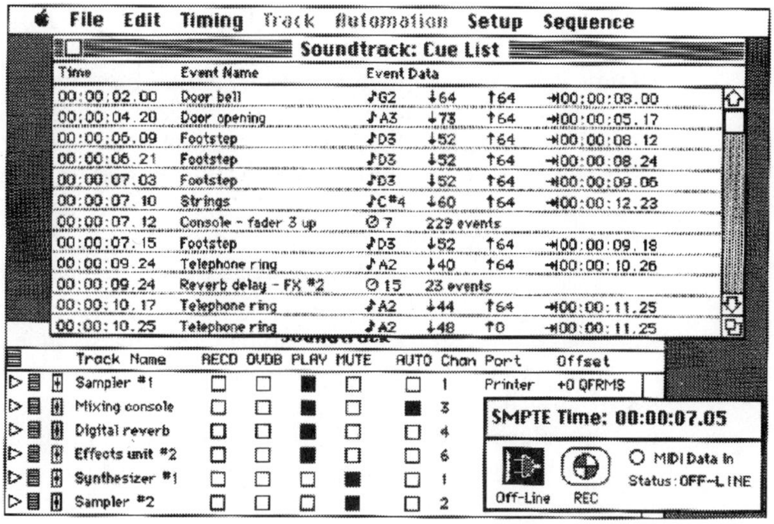

Fig. 5-22. Q-Sheet A/V cue-list display. (*Courtesy of Digidesign, Inc.*)

Q-Sheet A/V is much like a music sequencer. It records, stores, and plays back different types of MIDI events using individual tracks. Although music sequencers combine multiple tracks to assemble a song, Q-Sheet A/V combines one or more tracks to create a user-created cue list of MIDI events. An event might be placed into this list as the real-time movement of a fader, the setup of an effects parameter using SysEx messages, the adjustment of a delay setting, or, simply, a note-on/off command for a sampler or synth.

In addition to MIDI trigger/controller messages, Q-Sheet A/V offers integration features with Digidesign's Sound Tools Digital Recording and Editing System. This feature allows two independent tracks of hard-disk-based digital audio to be triggered in real time, along with simultaneous MIDI events. Once the music composition and/or visual sound track has been created, the entire cue list can be used to trigger and automate a mix under time-code control with quarter-frame accuracy.

The Algorithmic Composition Program

Algorithmic composition programs are computer-based interactive sequencers that directly interface with MIDI controllers or standard MIDI files to internally generate MIDI performance data according to a computer algorithm. This generated musical data can be used for gaining new ideas for a song, automatic accompaniment, improvisational exercises, special performances, or just plain fun.

This type of sequencer makes use of several methods for program/performer interaction. Initial program parameters can be entered by the user to control the performance according to musical key, notes to be generated, basic order, chords, tempo, etc. Alternatively, these performance guidelines can be varied in real time from the computer keyboard, mouse movements, or MIDI data.

Jam Factory from Intelligent Music (Fig. 5-23) is one such real-time interactive composing and performing program for the Macintosh computer. Once the user has played musical material from a MIDI keyboard or external sequencer, the program allows up to four automated players to learn and improvise around this composition. Each player's improvisations can be tailored to create variations on a musical passage or range of musical parameters over a number of playing styles.

Jam Factory allows the user to store and recall particular player-control settings and to record real-time parameter changes which effect an improvisation. It offers MIDI sync for synchronization with drum machines and other devices, as well as MIDI file compatibility with other Macintosh music software. The latter allows MIDI files to be used as source material for the automated players or for processed compositions to be exported to a sequencer.

M 2.0 is a real-time interactive composing and performing program from Intelligent Music. Unlike a sequencer, M's graphic-screen controls (Fig. 5-24) permit a user to shape or change any aspect of a MIDI-based composition while hearing it. The program allows the user to first specify which notes and chords

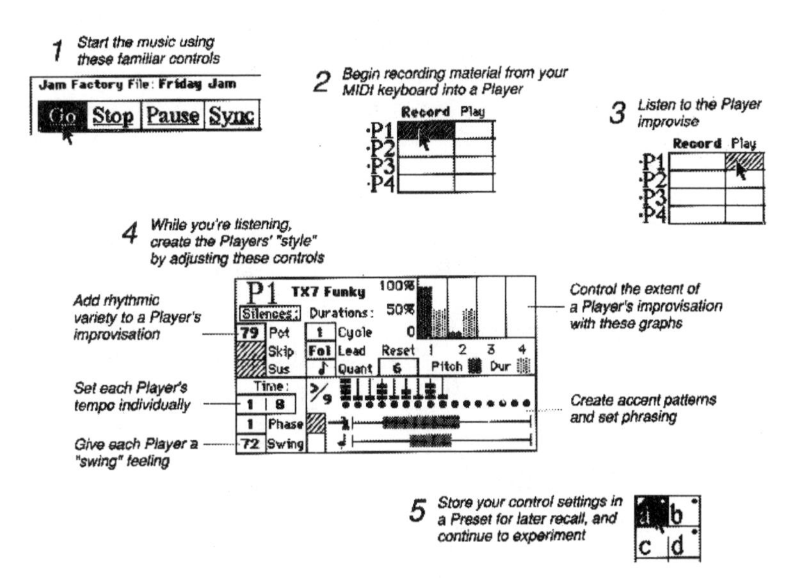

(A) Method of operation.

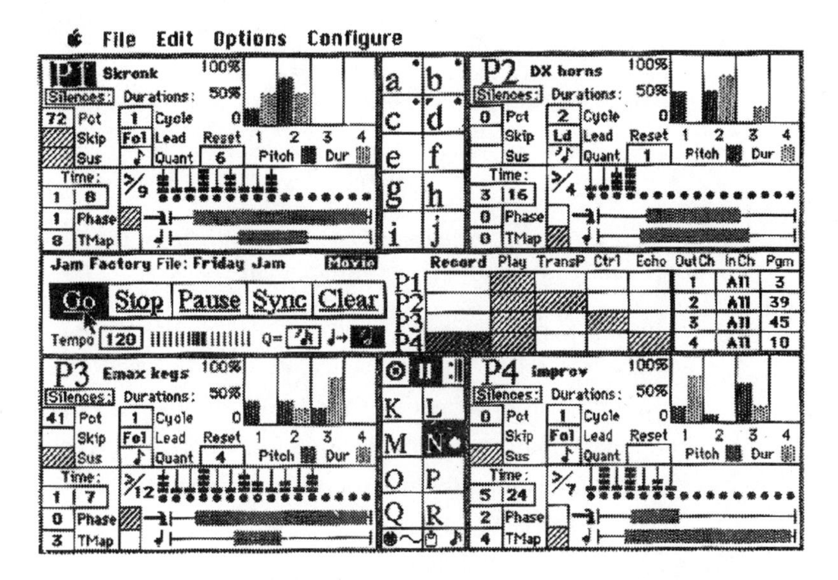

(B) Principal Jam Factory graphics screen.

Fig. 5-23. Jam Factory interactive composing and performing program. *(Courtesy of Intelligent Music)*

are to be played, then to determine the way those notes and chords will be transformed through rhythms, articulation, orchestration, and many other variables. Finally, the performance can be created and recorded, either by manipulating screen controls, playing-control keys on a MIDI keyboard, moving the mouse in a conducting grid, or creating automatic performance processes.

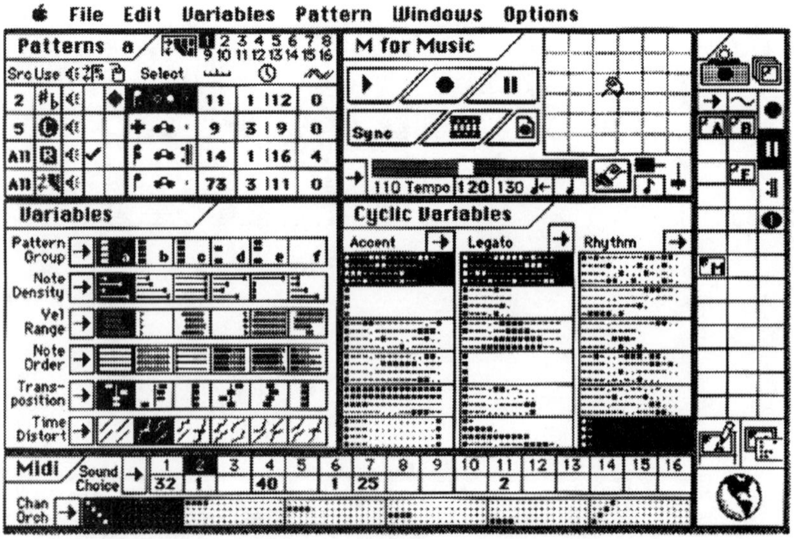

Fig. 5-24. Principal M graphic screen for the Macintosh computer. (*Courtesy of Intelligent Music*)

With programs available for the Macintosh, Atari ST or Mega ST, and the Amiga family of personal computers, M also offers:

- The capability for automating changes in MIDI velocities, note densities, rhythms, legato-staccato articulations, and accents.
- A time distortion feature for adding a human feel to the music.
- A snapshot feature which allows for storage and recall of control settings.
- MIDI file compatibility with other performance and notation software.
- MIDI sync capabilities.

Sound Globs from Cool Shoes Software (Fig. 5-25), is an interactive, compositional software package for the IBM/compatible computer. This program offers two distinct, but related, environments in which to work. The edit page allows various sound textures to be created and offers control over all basic parameters, in addition to generating performance-probability distributions. Graphic windows allow sound textures to be modified in real time, and for the viewing of pitches over time in a piano-roll fashion. The performance page is designed for real-time improvisation and performance, either from an external MIDI controller or computer keyboard. Additionally, slider controls allow performance parameters to be varied either individually or in a group fashion for simultaneous control over multiple parameters.

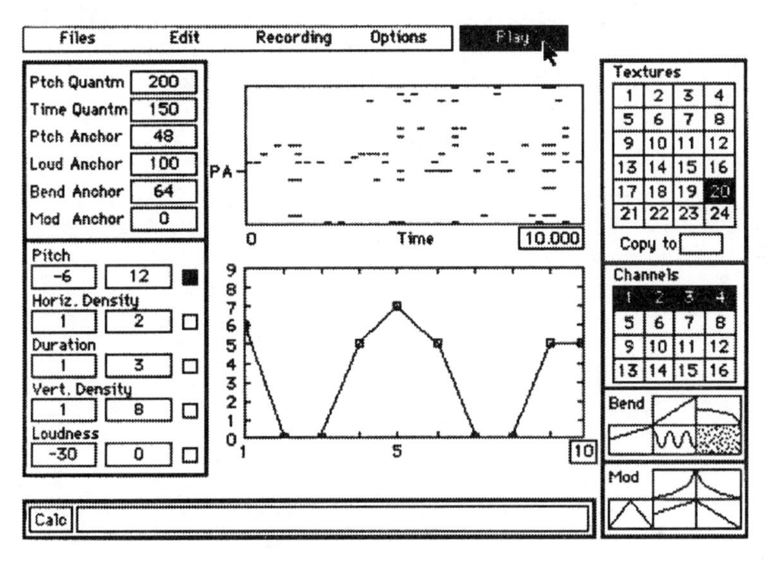

(A) Edit page screen.

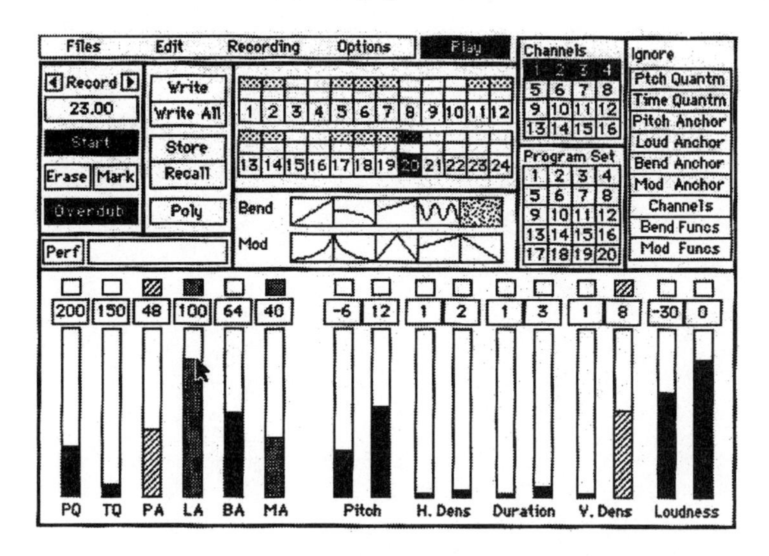

(B) Performance page screen.

Fig. 5-25. Sound Globs interactive musical sequencer.
(Courtesy of Cool Shoes Software)

Summary

As we have seen, both hardware and software-based sequencers are available in a wide range of configurations and include various features for the recording, editing, and reproduction of a musical performance. Each sequencer type (and for that matter, each model) has inherent design advantages and disadvantages, which may or may not best suit personal production needs. For these reasons, when looking for a sequencer, it is often a good idea to take the time to research or even test-drive various sequencer types, so as to find the one that best fits personal production needs.

Chapter 6

MIDI-Based Editor/Librarians

As we have seen from Chapter 4, most MIDI instruments and devices make use of a limited amount of internal RAM memory for storing internal patch data. Within a synthesizer, this memory contains data for controlling oscillators, amplifiers, filters, tuning, and other presets for creating a particular sound timbre or effect. In addition to the synthesizer, other MIDI devices often use memory locations for storing setup data which relate to performance parameters, such as effects settings, keyboard splits, MIDI channel and controller assignments, etc. These parameters may be manually edited by the user from the device's front panel and stored in an available memory location. Thereafter, patch and/or setup data can be instantly accessed by recalling the patch number and/or name from a bank of preset buttons, alpha dial, or keypad (Fig. 6-1).

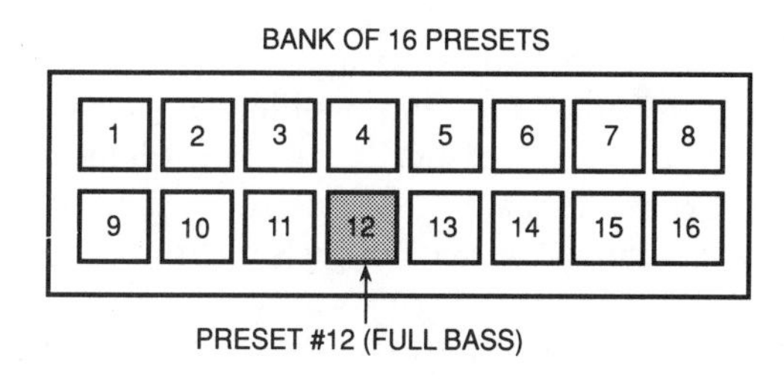

Fig. 6-1. Example of a bank of preset buttons for storing and recalling patch data.

MIDI devices are capable of managing preset memory data in a number of ways. One of the most basic forms of patch data is the *factory preset*. As the name implies, a factory preset is a bank of patches that have been programmed by the manufacturer to provide the performer with a set of initial sounds from which to build. This patch data may be encoded within ROM, making it impossible to edit and save new data within these locations. More commonly, factory presets are placed within battery-backed RAM, where they can be edited and rewritten over the original factory preset.

In the final analysis, factory patches are a collection of someone else's sounds, and may not reflect your own taste and style for sound texture. Many musicians, in fact, find that you're doing well, if you like even half of the sounds that are supplied with currently available instruments. This alone is incentive enough to set the most technologically timid on the road to editing and creating their own sound and parameter patches.

Patch data can be edited in a number of ways. One of the fastest and simplest ways is to alter the parameters of an already existing factory preset until a desired voicing effect or setup is achieved. By choosing a preset which is in the ballpark of the desired sound, half of the programming work will have been already done.

A second method is for the user to build a patch entirely from scratch. This way a unique voice structure can be constructed either by trial and error (by listening to the effect as parameters are changed) and/or by using your own experiences in synthesis and waveform analysis to build a patch. This may be accomplished by using the instrument's or device's front panel controls. Alternatively, a MIDI device may be edited through the use of a MIDI data-fader controller or through the use of a hardware- or software-based program editor. These enable the user to directly edit a compatible device by allowing control over MIDI parameters in real time.

The design of early analog synthesizers made use of a modular building-block approach. For example, an oscillator could be amplitude modulated by being patched into a *VCA* (*voltage controlled amplifier*) and then into a filtering module, etc. The control parameters for each of these sound producing or shaping modules physically consisted of rotary or slide faders and controls. This hands-on approach made for quick and easy access to each parameter. However, these controls required a large amount of space, were rather expensive, and made it difficult (if not impossible) to store settings for later recall.

With the advent of the digitally controlled analog and fully digital synthesizers, control over the many voice and setup parameters have been placed within the digital domain. This made it possible for the multitude of individual controls to be replaced with a single central-control panel (i.e., buttons, alpha dial, or keypad) that could be assigned to any of the wide range of digital control functions. This arrangement is basically the opposite of an analog system, since it is much less expensive and more compact. Unfortunately, however, this meant that the parameter controls, which were once easily accessible, are now imbedded deep within the programming architecture of the device's microprocessor. This often requires that the user tread through layers of control edit steps to change the digital parameters within a specific module.

The Patch Editor

An increasingly popular means for regaining real-time control over the control parameters of a MIDI device is through the use of a computer-based patch editor. A *patch editor* is a software-based package that is used to provide direct control over a compatible MIDI device, while clearly displaying each parameter setting on the monitor of a personal computer. Direct communication between the software/computer system and the parameter controls of a device's microprocessor is most commonly accomplished through the transmission and reception of real-time MIDI SysEx messages (Fig. 6-2).

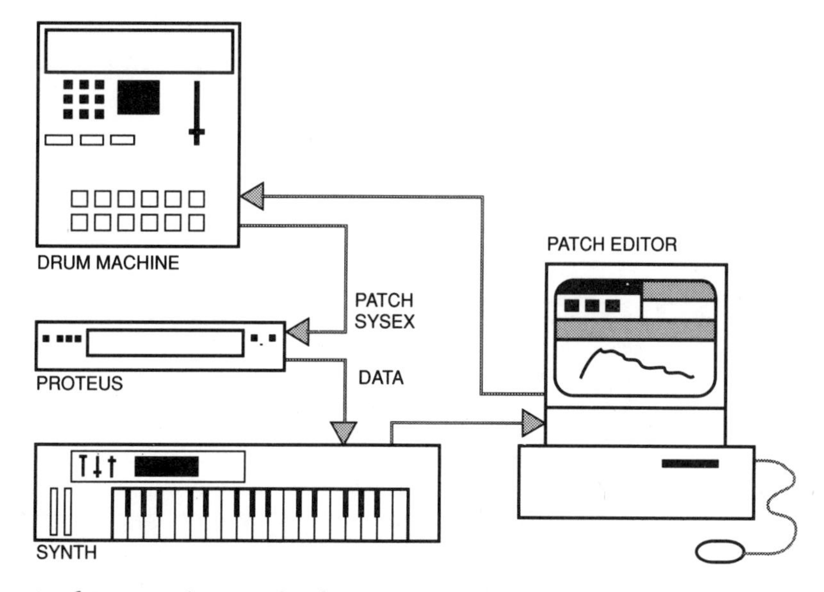

Fig. 6-2. A simple example of SysEx data distribution within a MIDI system.

The way patch editors are graphically laid out on a computer monitor will often vary widely from one software manufacturer to the next. This also holds true between MIDI device types, often depending upon the function and type of parameters that are to be edited upon the device. Many editors are designed to interact with a MIDI device using a numeric interface. That is, each of the various parameters to be controlled can be varied by changing the numeric value of that parameter on the computer screen. Often these values can be changed from the computer's keyboard or by the movement of a mouse (Fig. 6-3). In addition to numeric control, many programs graphically display waveform envelope data or other parameter curves directly upon the screen (Fig. 6-4). Such a screen display allows the user to instantly visualize important parameters, while often letting him or her edit them directly by placing a mouse cursor over the graph and dragging it to a new setting.

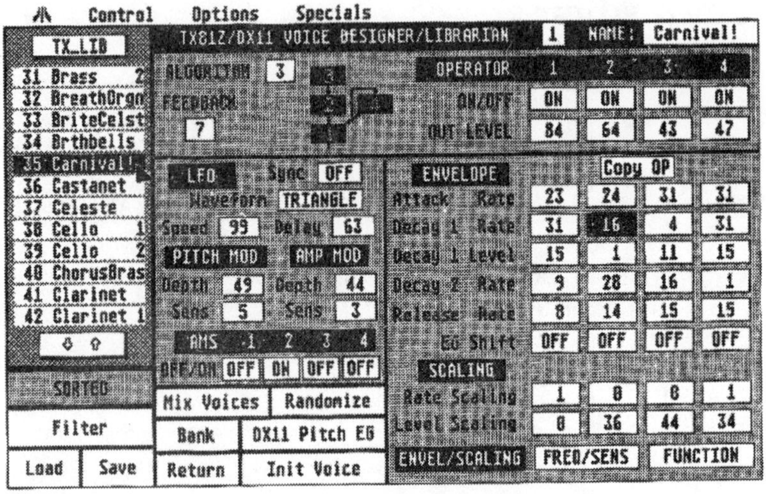

Fig. 6-3. The TX81Z/DX11 Voice Development System for the Atari ST. (*Courtesy of Musicode*)

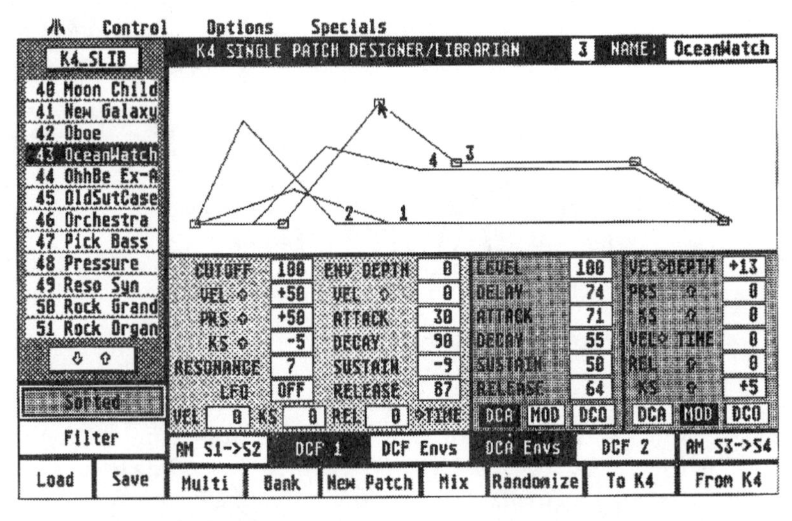

Fig. 6-4. The K4 Voice Development System for the Atari ST. (*Courtesy of Musicode*)

Given the fact that the many newer MIDI devices incorporate a large number of programming variables into their voice architecture, many voice and setup editors provide a number of tools for simplifying the editing process. This includes the ability to graphically zoom in on a range of related parameters, offer standard cut and paste capabilities for moving and/or copying a range of related parameters within and between patches, or for linking one or more parameters to an external controller (Fig. 6-5).

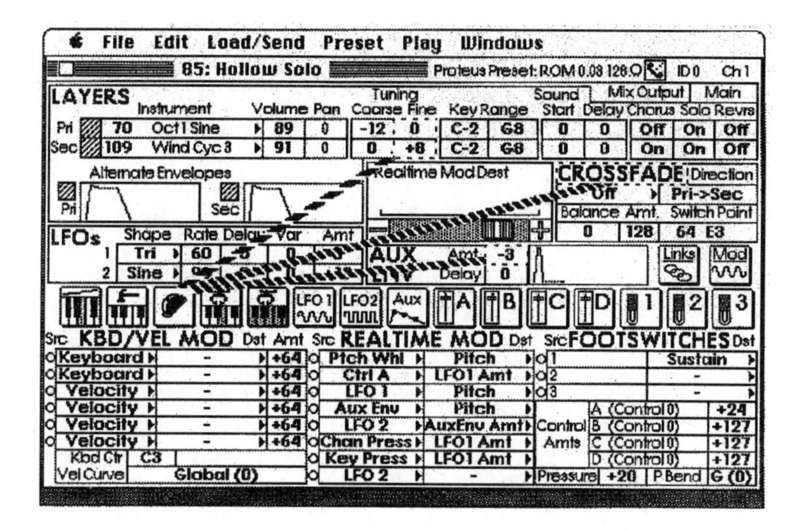

Fig. 6-5. Proteus editor/librarian showing linked parameters.
(*Courtesy of Opcode Systems, Inc.*)

Almost all popular voice and setup-editing packages include provisions for receiving and transmitting bulk patch data between the computer and MIDI device. This facility makes it possible to save and organize large amounts of patch-data files to floppy or hard disk. Many editing programs also provide options for printing out hard-copy versions of patch data.

Universal Patch Editor

In addition to dedicated patch editors, which are exclusively programmed for use with a specific instrument or MIDI device, it is becoming increasingly common to encounter programs which are able to communicate with a wide range of MIDI instruments and effects devices. These programs, knows as *universal editors*, are designed to receive and transmit device-specific SysEx data to provide on-screen control over the programming functions of most (if not every) MIDI device within a system.

Understandably, it is a difficult task to graphically represent the various faders, knobs, and other parameter controls for every possible MIDI instrument. As with single-device editor programs, many universal editors provide a numeric interface for altering patch data in real time. In addition to numeric parameters, many universal editors make use of a number of customized templates (an on-screen graphic layout of the controlled device), which are automatically invoked upon selecting the instrument or device. Finally, a few graphically oriented universal editors for the Macintosh and Atari family of computers allow the user to design their own customized template for replicating front panel or other relevant programming controls.

One universal editor available for the Atari ST, IBM/compatible, and Macintosh family of computers is X-OR Version 1.1 from Dr. T's Music Software (Fig. 6-6). X-OR is a universal SysEx editor/librarian system capable of receiving, transmitting, loading, and saving individual patches or entire banks from any MIDI instrument within a system.

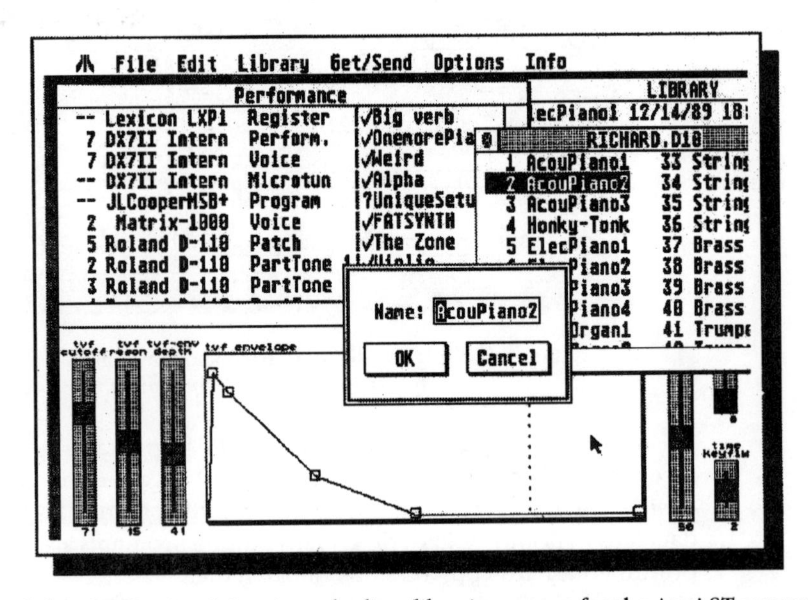

Fig. 6-6. X-OR Version 1.1 universal editor/librarian screen for the Atari ST computer. *(Courtesy of Dr. T's Music Software)*

X-OR includes a wide range of instrument profile and setup screens that contain all the necessary control parameters which are needed to establish SysEx communication. These profiles provide the user with on-screen parameter sliders, buttons, and graphic-envelope editing for each instrument. Also included are advanced features, such as four types of intelligent patch randomization, MIDI merging with solo and rechannelization modes, mouse play, and multiple banks in memory.

Another universal editor is GenEdit from Hybrid Arts, Inc. (Fig. 6-7). Designed for the Atari ST and Macintosh computers, GenEdit is a universal editor, librarian, organizer, and controller that is capable of working with any device with SysEx capabilities. However, a wide range of control-panel templates are supplied with the program, a built-in template editor permits the user to create a front panel or custom set of controls, using a straightforward paint applications program.

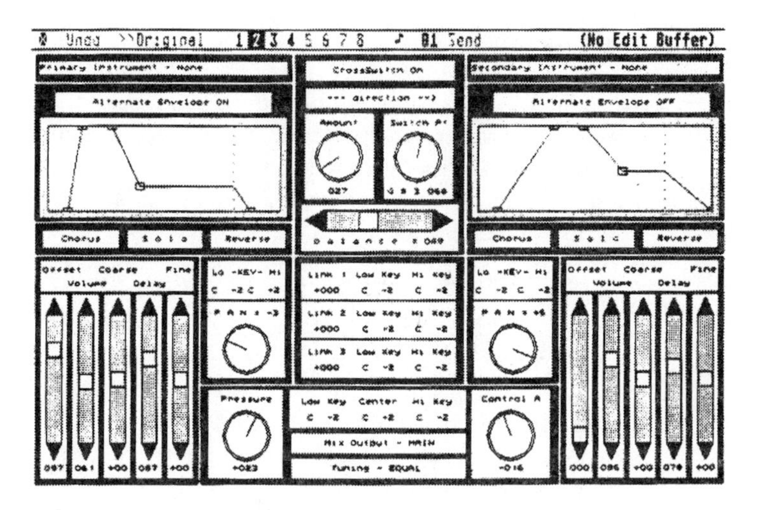

Fig. 6-7. The Proteus template from the GenEdit universal editor. *(Courtesy of Hybrid Arts, Inc.)*

GenEdit's librarian features allow the user to organize and save multiple patch data to computer disk. Whenever possible, it permits up to 1,024 patches to be grouped within a single file. New patches may be generated by this program, either randomly or according to basic parameters and incoming MIDI messages can be displayed through the use of the program's MIDI monitor screen.

The Patch Librarian

After having reedited currently existing factory patches or creating new patches from scratch, the programmer will invariably run out of preset storage locations. Thus, the problem of finding a means of storing additional patch data, as well as arranging these patches into organized patch banks, will quickly surface.

Most digitally based MIDI systems provide a platform for communicating bulk patch data to an external source, such as another device or MIDI-equipped computer. A software-based utility, known as a *patch librarian*, makes use of this function by receiving and transmitting patch data between one or more devices and a personal computer system. As with a patch editor, a librarian is used to download and upload entire banks of data via SysEx messages. This allows a virtually unlimited number of patches to be saved onto floppy- or hard-disk drives. Such a system provides us with an easily managed, cost-effective system for storing data.

Patch librarians also provide the user with a means for copying and moving patch data between banks or into new patch banks. This way the user can reorganize patches into any bank, according to sound type (i.e., bass, strings, effects, etc.) or any other desired criteria.

Universal Patch Librarian

As with the universal patch-editor program, it is becoming increasingly common for electronic musicians to make use of computer software that can communicate SysEx patch-bank data with a wide range of MIDI instruments and devices. Such a program is known as a ***universal patch librarian***.

As noted in Chapter 2, SysEx messages are used to communicate customized MIDI messages (such as patch parameter data) between MIDI devices. The format for transmitting these messages includes a SysEx status header, manufacturer's ID number, any number of SysEx data bytes, and an EOX byte. It is the role of a universal patch librarian to both record and distribute this data in a format that allows patch data to be individually communicated to the appropriate devices within a MIDI network.

Sequencer packages that integrate a universal patch librarian into their program package are often capable of automatically transmitting appropriate patch data, setup, and program-change messages to each device that is within a MIDI system. In this way, a librarian can be used to automatically reconfigure the voice and setup patches throughout a system in advance of playing the sequence. Certain systems allow for such setups to be directly tied to a song file or given a separate name for use as a performance template. Thus, such a template could be used to provide system patch, routing, and setup data within a live or studio performance setting. For example, Figure 6-8 shows a setup page in which up to 32 MIDI devices can be programmed into a single setup page (over 100 MIDI devices are currently supported by the software). This setup page contains the respective patch banks, MIDI channel, and program names for every device within the system's overall snapshot. Upon initiating the transmit command, the selected patch banks are sequentially transmitted to each instrument and/or device via MIDI SysEx dump. Once these devices are loaded, the proper program-change commands are then transmitted, often only requiring the user to press the play button to hear the song with the correct voicings and settings.

Another universal librarian package is Opcode's Galaxy (Fig. 6-9) for the Macintosh computer. Galaxy is capable of supporting over 70 MIDI devices, and is directly compatible with files that have been created with any of Opcode's other editor/librarian packages. In addition, a ***Patch Talk*** language has been created, which enables the user to design file types for future instruments.

Many newer MIDI instruments make use of multiple banks, which are used to store complex voice and setup parameters. When combined, they create a range of voices which might include such catagories as patches, timbres, rhythm setups, etc. Although these patch categories contain data relating to a single instrument, they are often stored and named as separate banks. Galaxy is capable of grouping these related data banks (along with the banks of other instruments)

Librarian

Setup TAKE 1 Mem 218496
Instrument KORG M-1 Prog Link

Instrument	Chan	Mode	Bank	Prog#	Program
YAMAHA DX7/TX	1	POLY	DX5B-2	0	E.P.& BR B
KORG M-1	3	POLY	SET 1	2	Brass 1
KORG M-1	4	POLY	SET 2	62	TENOR SAX
YAMAHA TX81Z	5	POLY	TEST VER	5	ELECTRIC GUITAR
VOYETRA 8	6	POLY	EFFECTS	12	EXPLOSION
KORG DW/EX8000	7	POLY	SPACE	3	SAX FROM SPACE
ROLAND JX-8P	9	POLY	ORCHESTR	42	STRINGZ ZZ
———	1	POLY	———	0	———
———	1	POLY	———	0	———
———	1	POLY	———	0	———
———	1	POLY	———	0	———
———	1	POLY	———	0	———
———	1	POLY	———	0	———
———	1	POLY	———	0	———
———	1	POLY	———	0	———
———	1	POLY	———	0	———

Librarian Menu
Instrument Bank Program Transmit Single Delete Link-prog Report
Open-mem ARRANGER MIDI-TERMINAL FILES

Fig. 6-8. Setup page from Voyetra's network organizer/universal librarian program for the IBM/compatible computer. (*Courtesy of Voyetra Technologies*)

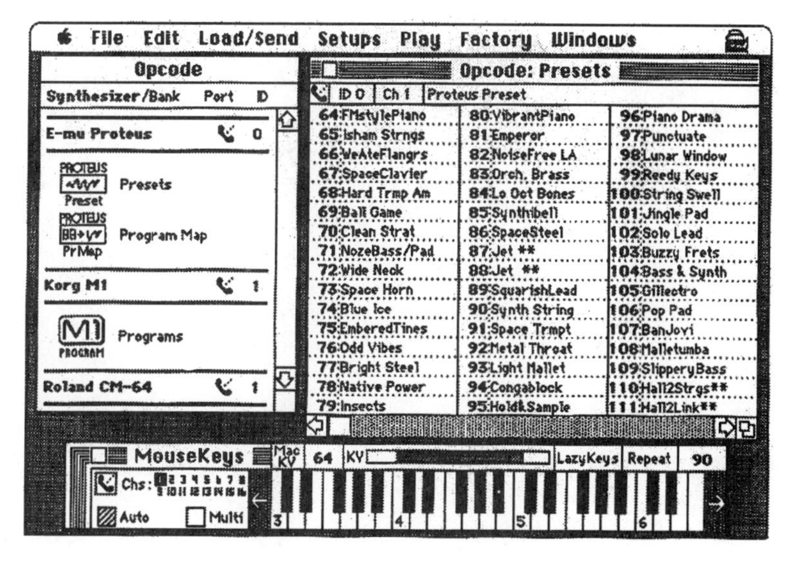

Fig. 6-9. The main screen of the Galaxy universal librarian. (*Courtesy of Opcode Systems, Inc.*)

into an overall snapshot known as a *bundle* (Fig. 6-10). This approach makes it easier to visually determine which patch groupings are directly related to their instrument, and to simplify an otherwise complex patch-management arrangement.

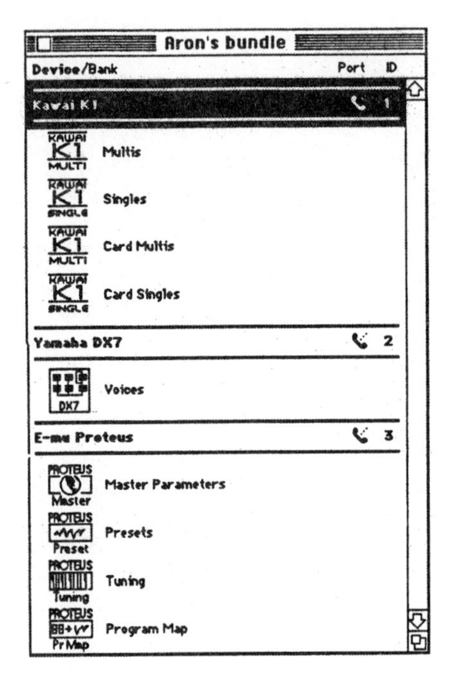

Fig. 6-10. A patch bundle from the Galaxy universal librarian. (*Courtesy of Opcode Systems, Inc.*)

Through the use of Apple's computer-program switching and multitasking utilities (such as Apple's MultiFinder and MIDI Management Tools Set), Galaxy is capable of linking each patch name within the library to its professional sequencer, Vision. This allows the user to directly select patches from within the sequencer by name (not number). In addition, anytime a change is made in Galaxy that corresponds to a selection or setting in the sequencer, this change is automatically copied to the other program.

Alternative Sources for Obtaining Patch Data

In addition to factory and user-created custom patches, other methods exist for obtaining patch data from within the commercial and public domain.

Preprogrammed patches are available for almost every popular MIDI instrument and effects device, spanning almost every timbre and instrumentation style. These patch banks have been programmed by working professionals and/or dedicated patch heads using a wide range of formats, including computer diskette, (for loading patches directly into a MIDI device or dedicated librarian package), ROM/RAM card, and data cassette. These professional and homegrown products can be commonly found at a reasonable cost in the classified section of most magazines which cater to the electronic musician.

In addition to the previous means for storing patch data, numerous books are available which contain hard copies of patch-parameter values. Using this method, the user must manually enter each patch parameter value into the MIDI device. Although this process will often take a great deal of time and effort, these low-cost publications allow a large number of patches to be distributed without the complications of providing patch data in a wide range of computer and librarian data formats.

In addition to distribution media, banks of patch data exist within the various computer bulletin boards (BBS) that provide a platform for downloading and uploading MIDI-based computer files. Bulletin boards (such as Pan and Compuserve) provide a range of patch databases and related MIDI programs for most computer types and popular librarian packages. These services also provide access to public domain programs which have been custom designed by BBS users. Additionally, they include dedicated MIDI editors, librarians, SysEx applications, and a multitude of other MIDI program types.

Chapter 7

Music-Printing Programs

In recent years, the field of transcribing musical scores onto paper has been directly affected by both computer and MIDI technology. This process has been enhanced through the use of newer generations of computer software which commonly allows musical notation data to be entered into the computer either manually or via MIDI. Once entered, these notes can be edited in an onscreen computer environment which enables the artist to change and configure a musical score or lead sheet using standard cut, copy-and-paste techniques. In addition, many programs allow for a score to be played back by transmitting MIDI messages to various connected electronic instruments within a system. A final and important capability of such programs is their ability to print out hard copies of a score or modern lead sheet in a number of desired print formats and styles.

Entering Music Data

Music-printing programs allow the musician to enter musical data into a computerized score using a number of input methods. Most commonly, for example, a *music-notation program* is designed to provide the user with a straightforward method for manually entering music notation into a scoring program. Many programs of this type offer a wide range of notation symbols and type styles, that can commonly be entered either from a computer keyboard or mouse. In addition to the manual entry of a score, a music-transcription program is capable of automatically entering a score into a computer through the use of MIDI. This may commonly be accomplished in real time (by playing a MIDI instrument or finished sequence into the program), step time (thereby notating a score one note at a time by playing a MIDI keyboard controller), or from a standard MIDI file (which uses an actual computer-sequence file as a source for MIDI data).

Both notation and transcription programs often vary widely in their capabilities, speed of operation, and number of features that they offer to the user. This includes such capabilities as: the number of staves which can be entered into a single score, the overall selection of available musical symbols, or the entering and editing of written text within a score.

Certain packages limit notation to a single grand staff, which is composed of an upper treble clef and lower bass clef. Other more comprehensive programs permit data entry onto a larger number of staves for compositional and orchestral notation. A comprehensive selection of musical symbols is commonly available when using such a program. Often a standard notation palette will include standard note lengths, rest duration markings, accidental markings (flat, sharp, and natural), dynamic markings (i.e., pp, mp, mf, and ff), and a host of other important score markings.

In addition to standard notation entry, it is possible to enter text into a score for the adding of song lyrics, song titles, special performance markings, and additional header/footer information. Text input and editing can often be easily edited within a score, allowing lyrics to be cut, copied, and pasted. Often a music-printing program is able to link the text of a score to the music notation itself, so that when the music is moved and/or justified into a final format for printing, the text is likewise automatically positioned to its new corresponding measure.

Commonly, a program can be set to specific default settings. This enables the user to change such parameters as measure widths and individual or global control over stem direction.

Editing a Score

Music-printing programs incorporate a system for editing a score which has been entered into it either manually or via MIDI. This is accomplished by allowing the user to add, delete, or change individual notes, durations, and markings by using a combination of computer-keyboard commands, mouse commands, or remote MIDI keyboard commands. Larger blocks of music data can also be edited using standard cut copy-and-paste methods.

One of the major drawbacks to automatically entering a score via MIDI (either as a real-time performance or by automatic entry from a standard MIDI file) is the fact that music notation is an interpretive art. "To err is human," and it is commonly this human feel which gives music its full range of expression. It is very difficult, however, for computers to interpret these imperfections. For example, they might interpret a held quarter-note as either a dotted quarter-note or one that is tied to a thirty-second note, etc. Although computer algorithms are getting better at interpreting musical data, and quantization can be employed to instruct a computer to *round* a note value to a specified length, a score will still often need to be manually edited by the user to correct misinterpretations that might be made by the computer.

Playing Back a Score

In the not-so-distant past, musicians commonly had to wait months and/or years to hear a finished composition. Orchestras and ensembles were generally expensive and often required the musician to have a good knowledge of politics. One solution to this was the piano reduction, which served to condense a score down into a compromised version that was playable at the piano keyboard.

With the advent of MIDI, it is possible for musicians to use the score for their newly-created composition as the source for transmitting MIDI messages out to predefined MIDI channels and devices. In this way, a system setup that reproduces the intended score, which is faithful or a close working approximation of the final performance, can be created with relative ease. The ability to playback a composition also provides the artist with a first hand means of checking for final errors within the score.

Printing a Score

Once the score has been edited into a final form, it is possible to create a hard-copy print. Generally, a notation program will allow the score to be laid out in a manner that best suits the user's taste or final application. Often a comprehensive program will allow parameters, such as margins, measure widths, etc. to be user-adjusted. Additionally, title, copyright, and other information may be added within the text mode. Once completed, the final score can be printed to either a dot-matrix, ink-jet, or laser printer (depending upon which printer type and font is supported by the software). A 9-, 12-, and 24-pin printer is capable of creating printed scores that range from acceptable to near-publishing quality. Many 300-dpi (dots-per-inch) ink-jet printers and laser-printing systems are capable of generating printed copies that are of publishing quality.

One example of a music-notation software program for the Macintosh family of computers is MusicProse from Coda Music Software (Fig. 7-1). This program has been especially designed for the creation of lead sheets and piano-vocal or small ensemble scores. It is capable of notating up to eight staves with up to four parts per staff. MusicProse offers up to four ways to input music:

- *Simple entry*: For use with step-time entry using the mouse and onscreen tool palette.
- *Speedy entry*: Step-time entry using the computer's keyboard and/or MIDI keyboard.
- *Hyperscribe*: Permits on-screen transcription from a MIDI instrument in real time.
- *Transcribe MIDI files*: Reads external sequencer files that support the standard MIDI file format.

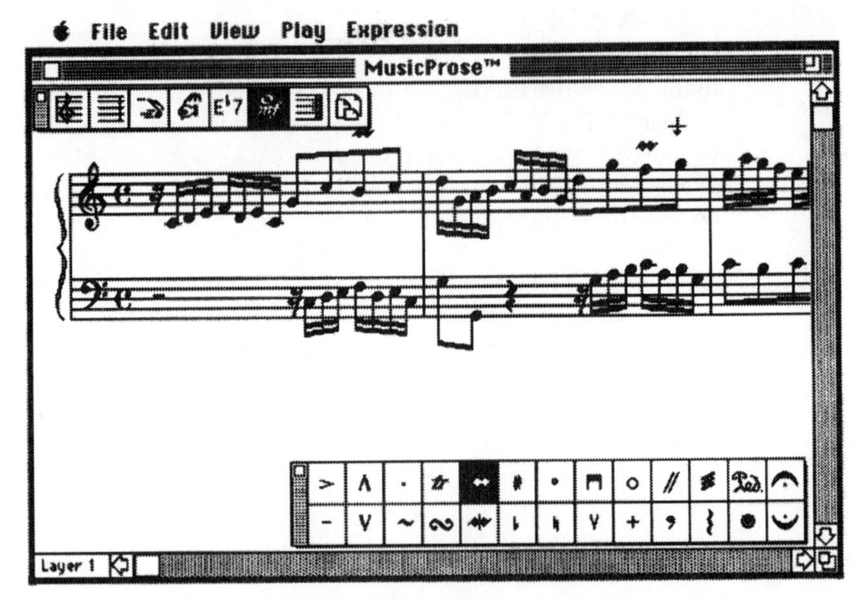

Fig. 7-1. MusicProse music-notation software program expression screen. (*Courtesy of Coda Music Software*)

After being entered into the program, a full range of edit parameters can be performed upon the score using mouse movements (i.e., move, add, replace, or delete techniques). Lyrics can also be entered into an arrangement and edited into the score in a number of ways that include a number of user-selectable fonts. Block chords and guitar fingerboard symbols can also be analyzed by the program and automatically placed into the score.

Once completed, a score can be played back via MIDI or directly from the computer's speaker in 4-voice polyphony. Printing is accomplished using Apple's LaserWriter printers and two PostScript fonts:

- *Petrucci*: Comprised of a complete set of traditional music symbols and an extended set of contemporary symbols.
- *Seville*: A guitar chord font and symbol library.

The MusicPrinter Plus Version 3.0 from Temporal Acuity Products, Inc. is a music-notation and MIDI performance program for the IBM/compatible and C1 computer. This program allows the user to enter an on-screen score page via computer keyboard, direct MIDI input (in either real or step time), or from a standard MIDI file. Entry from the keyboard is carried out by assigning a musical character to a specific computer key. Thus, the character is pasted onto the screen by placing the mouse at the proper point and pressing the desired key (i.e., q =

quarter note). When in the direct MIDI input mode, once a number of setup parameters are entered, a musical line can be notated by performing directly from a MIDI keyboard.

The MusicPrinter Plus allows the user to notate a score which contains up to 42 staffs with a full compliment of over 600 available text and musical characters. The word processing mode permits up to four text fonts to be entered into a score, in addition to foreign-language diacritical marks. Playback can be output directly from the scored notation, with a maximum of 128 simultaneous voices. During play, the score can be scrolled in an on-screen fashion (with possible control over tempo and early page turn), or can be stepped through one note at a time.

This program is capable of configuring a score for printing, with manual and automatic formatting of the computer screen's output into a printed page, system, and staff layout. Printing can be carried out using a number of 9-pin, 24-pin, and laser-printer configurations (Fig. 7-2).

Personal Composer 3.3 from Personal Composer is an integrated MIDI package for the IBM/compatible computer which includes music notation, a MIDI sequencer, MIDI editing capabilities, and music printing.

Music notation for a wide range of applications (including orchestral scoring with up to 64 staves) can be entered into Personal Composer from a PC's keyboard or mouse. Music can also be directly transcribed from a compatible MIDI file. (Textures and Sequencer Plus programs offer utilities for transferring sequencer files into the Personal Composer format.)

While scrolling along in the score, a MIDI representation of the music is placed into the 32-track MIDI recorder. When the score is played (Fig. 7-3) each staff can be sent to its own MIDI channel.

This program also contains a 32-track MIDI sequencing program (Fig. 7-4) that enables the user to record MIDI performance data which can be reproduced and/or used as a source for generating a musical score. This sequencer incorporates a wide range of editing commands in addition to extensive MIDI filtering capabilities for extracting user-defined MIDI parameters from a musical score. This sequencer can be synchronized internally to an external source or tape, and is capable of responding to SPP messages. A performance-controller feature is also included which enables songs to be accessed quickly and chained together, allowing a number of songs to be consecutively played in a live performance setting.

For those who are not adept at reading and manipulating music notation, a MIDI event editor (Fig. 7-5) is included which represents the musical score in a piano-roll fashion that closely resembles the edit window, which is included within most sequencer packages. Using this method, the pitch of a note is plotted vertically, while a note's length is plotted horizontally within the screen. A wide range of edit functions can be performed within this mode, including cut, copy-and-paste edit capabilities, graphically displayed quantization, note velocity, text marking, and a limited function for algorithmic composition.

Fig. 7-2. Various print examples from the MusicPrinter Plus Version 3.0 notation program. (*Courtesy of Temporal Acuity Products, Inc.*)

Fig. 7-3. Basic score-editing screen. (*Courtesy of Personal Composer*)

RECORDER	1	2	3	4	5	6	7	8	9	10	11	12	13	14	15	16
	1	2	3	4	5	6	7	8	9	10	11	12	13	14	15	16
GRAPHICS	•	•	•	•	•	•	•	•	•	•	•	•	•	•	•	•
MUTE																
L CHANNEL																
R CHANNEL																
L FADER	16	16	16	16	16	16	16	16	16	16	16	16	16	16	16	16
R FADER	16	16	16	16	16	16	16	16	16	16	16	16	16	16	16	16

	17	18	19	20	21	22	23	24	25	26	27	28	29	30	31	32
GRAPHICS	•	•	•	•	•	•	•	•	•	•	•	•	•	•	•	•
MUTE																
L CHANNEL																
R CHANNEL																
L FADER	16	16	16	16	16	16	16	16	16	16	16	16	16	16	16	16
R FADER	16	16	16	16	16	16	16	16	16	16	16	16	16	16	16	16

RECORD OFF
lead-in 4/4
track# 2
channel# 1
notes ON
control OFF
program ON
pressure OFF
pitchbend OFF

PLAY
FAST OFF
METRONOME OFF
LOOPING AUTOSTOP
1ST MEAS 1
LAST MEAS 200
BOUNCE
ERASE
SCORE

TIME SIG 4/4
TEMPO 100
MIDI THRU ON
SYNC INTERNAL
EXIT
ZAP
START AUTO
FOLLOW OFF

2 measures
42 bytes

BUFFERS
1 2

Fig. 7-4. MIDI recorder screen. *(Courtesy of Personal Composer)*

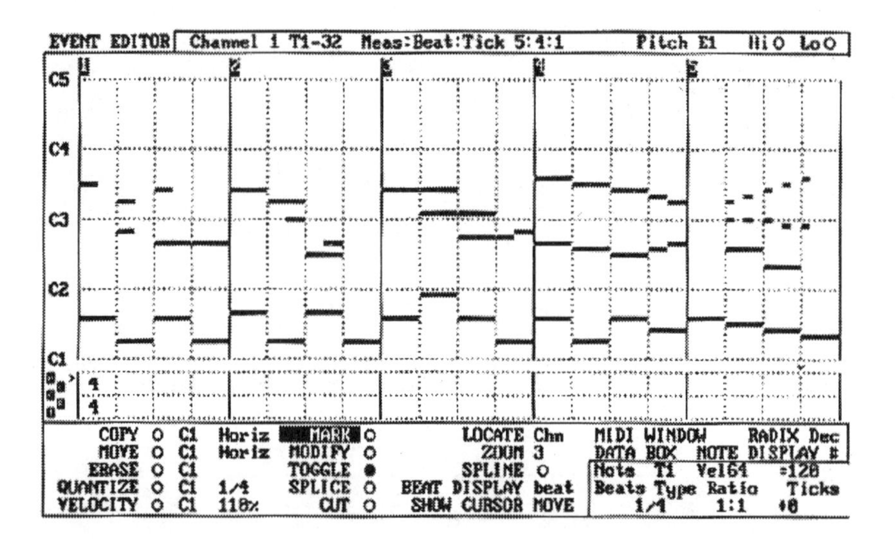

Fig. 7-5. Event editor screen. (*Courtesy of Personal Composer*)

Personal Composer also incorporates a synthesizer patch librarian and editor which is specifically designed for the Yamaha DX-7, TX-7, and TX-816 line of synthesizers (Fig. 7-6). In addition, this patch facility is capable of manually receiving and transmitting any form of SysEx data that does not require handshaking between the device and computer.

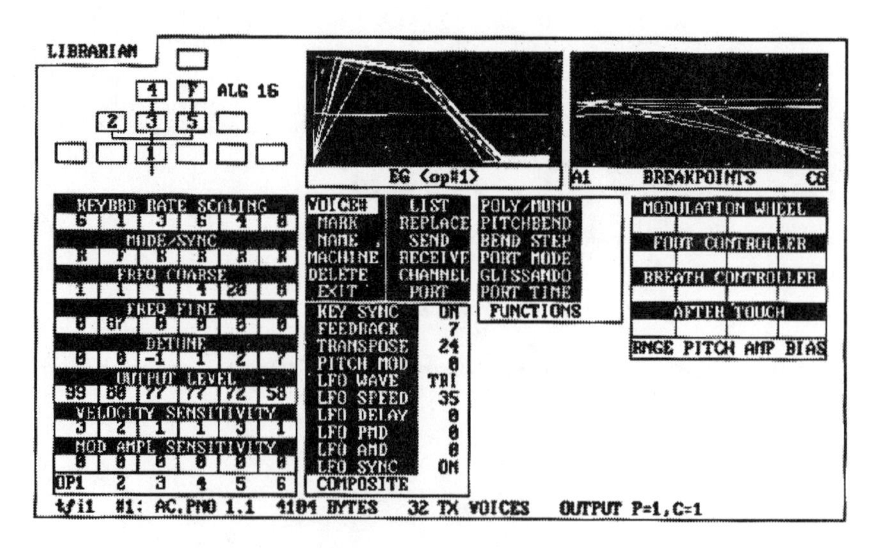

Fig. 7-6. Librarian screen example (for TX synthesizers). (*Courtesy of Personal Composer*)

Notation display and editing have also become a feature within other types of MIDI software packages. For example, Performer Version 3.4 has added notation editing (Fig. 7-7) of MIDI tracks to its standard events list and graphic-editing capabilities. Instead of a standard pitch ruler and note grid, the notation-editing window displays notes on a musical grand staff. This function also includes note cut, copy and move capabilities, a time ruler, markers strip, median strip, and a continuous data strip. An octave up/down selector is provided to graphically center any octave on the staff.

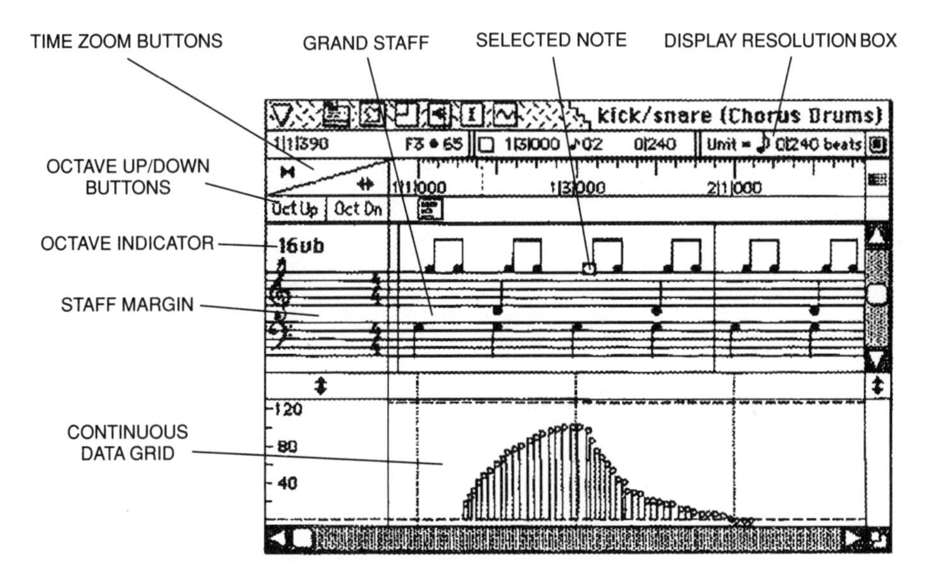

Fig. 7-7. The Performer Version 3.4 notation-editing window. *(Courtesy of Mark of the Unicorn)*

In conclusion, music-notation programs often vary in their editing, playback, and printout capabilities. They also vary in the number and type of features that are offered, and in the personal feel of the program. For these reasons, it is wise to shop carefully and compare programs which show promise of best suiting your creative needs.

Chapter 8

MIDI-Based Signal Processing

As within professional recording technology, MIDI production relies heavily upon the use of electronic signal processing for the re-creation of a natural ambience or for augmenting and modifying audio signals.

Modern effects devices offer a diverse range of effects and signal processing capabilities (such as reverb, delay, echo, auto pan, flanging, chorusing, equalization, and pitch change). These devices also offer a wide range of control parameters (such as delay time, depth, rate, filtering, etc.), which can be varied by the user to achieve a processing effect that best suits the program material. With the implementation of MIDI in many of today's digital signal processors, it is possible for precise and/or automated control over the above effects and their parameters to be integrated into a MIDI system (Fig. 8-1).

Effects Automation within MIDI Production

One of the most common methods for automating effects devices within a MIDI sequence or live performance is through the use of MIDI program-change commands. In the same manner that data relating to a characteristic sound patch can be stored within an instrument's memory-location register for later recall, most MIDI-equipped effects devices offer a number of memory registers in which effects patch data can be stored. Through the transmission of a MIDI program-change command over a particular MIDI channel, it is possible for such effects patches to be automatically recalled.

The use of program-change commands (and occasionally continuous-controller messages) allow complex signal processing functions to be easily modified or switched during the normal playback of a MIDI sequence. Often, a sequencer will allow the simple insertion of a program-change number (that corresponds to the desired effects patch) into a separate sequenced track, or onto a track that contains related performance data.

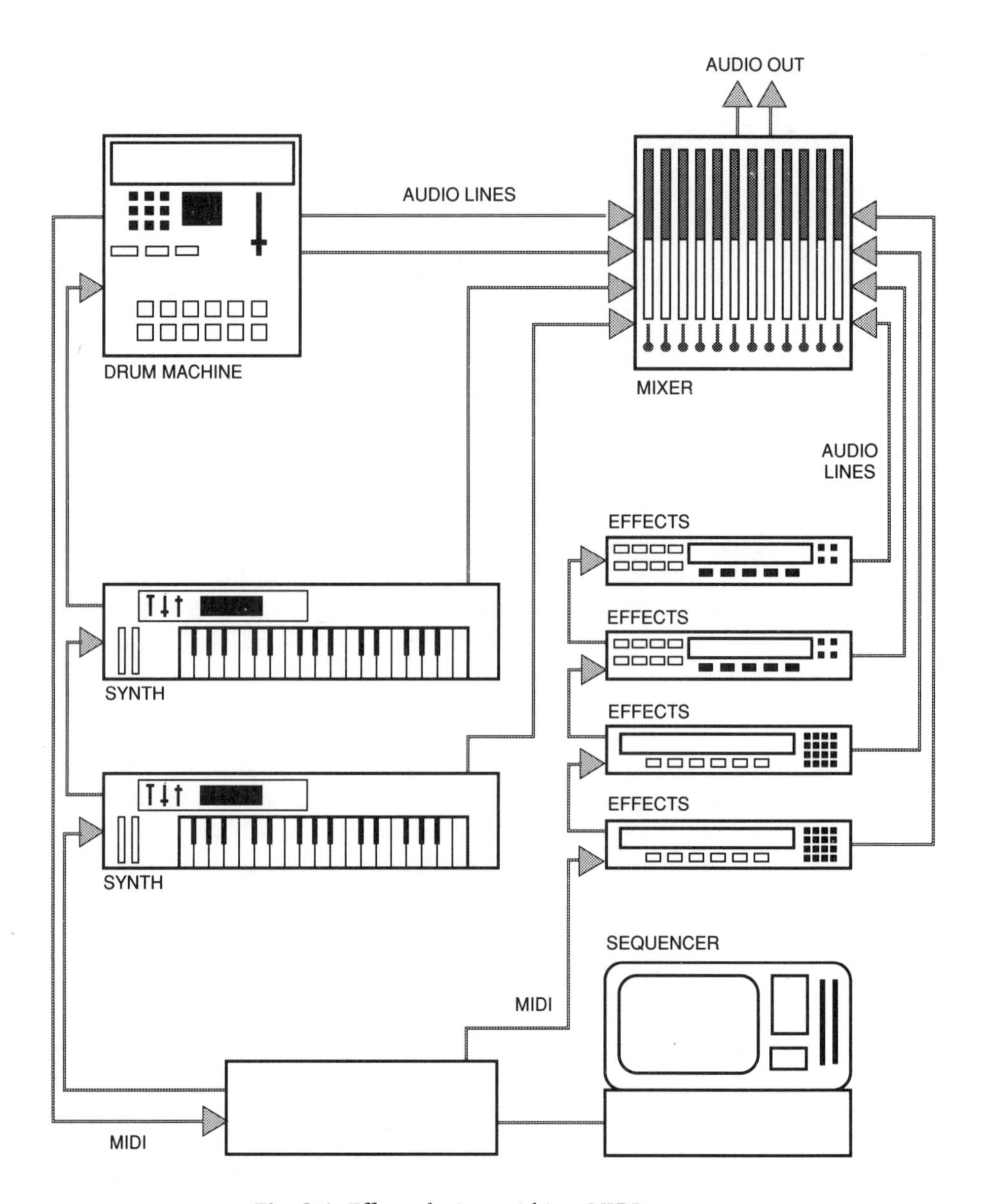

Fig. 8-1. Effects devices within a MIDI system.

Dynamic Effects Editing Via MIDI

In addition to real-time program changes, dynamic editing of effects parameters is often possible through real-time SysEx messages for direct control over preset effects parameters (i.e., program type, reverb time, equalization, or chorus depth). Control over these messages may often be accomplished in real time through the use of an external MIDI data-fader controller or hardware/software-based controllers (Fig. 8-2), which allow for direct control over effects parameters via the transmission of SysEx messages.

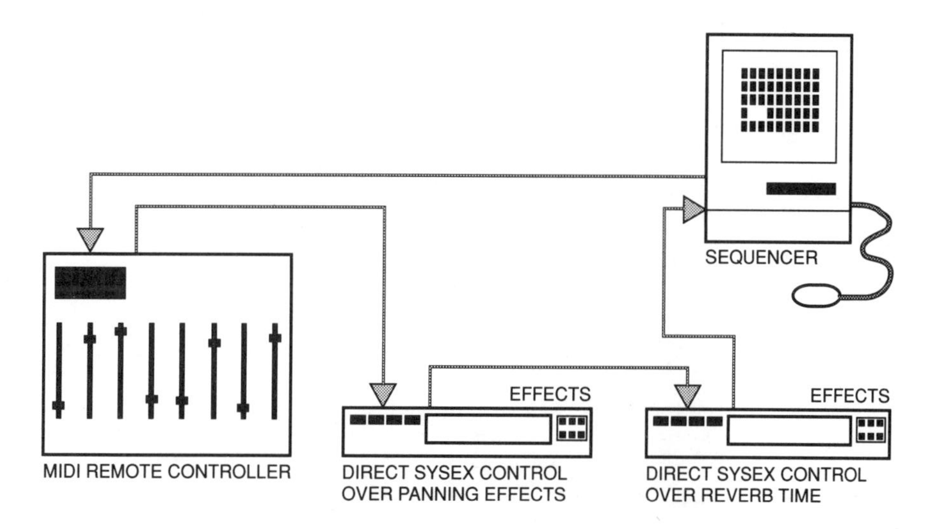

Fig. 8-2. Dynamic control over effects parameters via external MIDI remote controller.

Computer-based effects editing software (Fig. 8-3 and 8-4) also offer dynamic editing of effects parameters by allowing the user to edit and fine-tune effects parameters using onscreen mouse-driven visual graphics and scales. Another way is through the display of numeric values which directly represent the device's control parameter settings. Control over these parameters is accomplished in real time through the use of device-specific SysEx messages. Once the desired effect or multi-effects have been assembled and fine-tuned, these parameter settings can be saved to one of the device's preset registers for recall at a later time, either from the device's front panel or via program-change messages.

In the recent past, program editors have often been designed to operate with a specific device or limited range of devices. This is due to the fact that these editors are required to communicate SysEx data which pertains solely to the device being edited. With the large number of MIDI-controlled effects devices appearing on the market, it is becoming increasingly more common to find universal editors which are capable of directly editing a wide range of devices from various manufacturers in real time. As such, these universal programs must be capable of effectively managing a wide range of devices. Therefore, they often contain a more generic visual interface for controlling the effects devices and musical instruments than their dedicated counterparts.

Once a device's bank of preset locations has been filled, a program editor will often allow these SysEx messages to be transmitted to computer by way of a SysEx

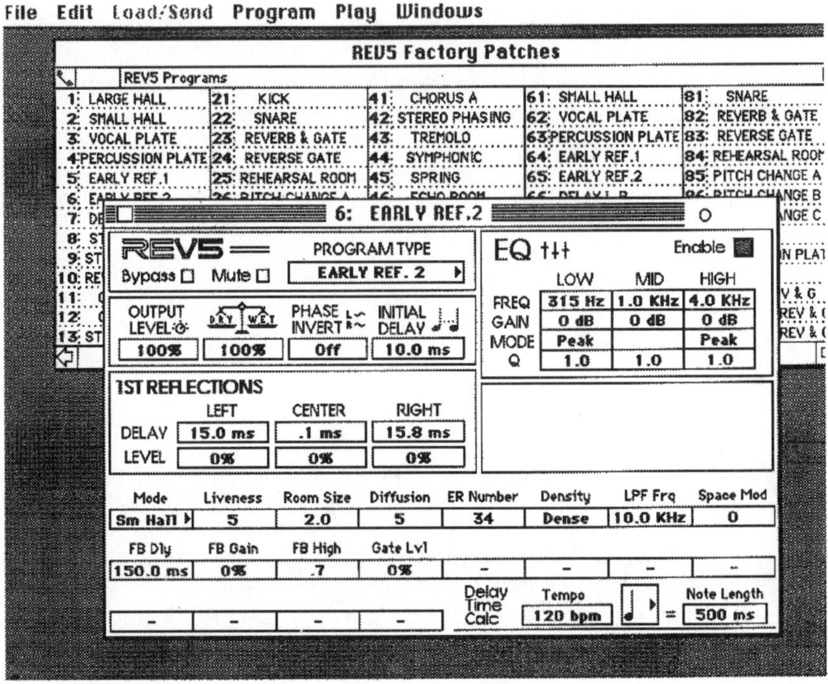

Fig. 8-3. Opcode REV-5 editor/librarian. (*Courtesy of Opcode System, Inc.*)

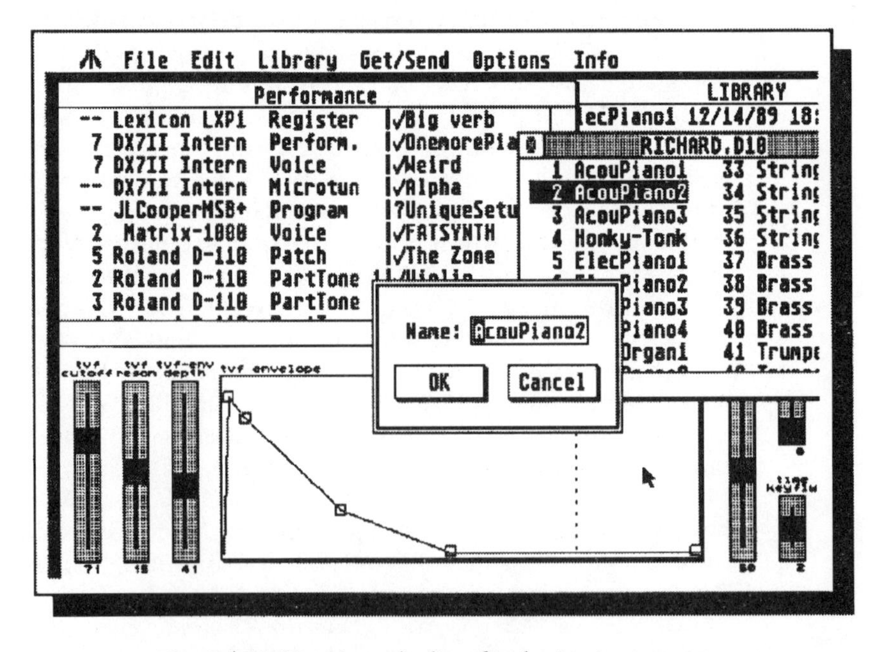

Fig. 8-4. X-OR universal editor for the Atari computer.
(*Courtesy of Dr. T's Music Software, Inc.*)

MIDI data dump. In this way multiple preset banks can be stored and recalled, allowing a much larger number of effects patches to be stored within a computer-based library. Such programs, commonly referred to as patch librarians, also permit the user to organize effects patches into groups of banks according to effects type or in any other order that the user desires.

As with most electronic instrument patches, effects patches can be acquired from a number of sources. Among these are specific patch books (containing written patch data for manual entry), patch data cards (ROM cards or cartridges from the manufacturers or third party developers containing patch data), data disks (computer files containing patch data from manufacturers or third party developers), and computer bulletin board files (containing patch data which can be downloaded via computer modem).

MIDI-Equipped Signal Processing and Effects Devices

Within recent years, the number of signal processing and effects devices that are implemented with MIDI has steadily increased. This is due to the power and increased flexibility that these devices bring to MIDI production, as well as the increased editing and control capabilities which MIDI is able to bring to the field of signal processing.

For the remainder of this chapter, we shall overview many of the various types of MIDI-equipped processors and effects devices that are available for use in MIDI, recording, audio-for-visual, and live-performance production.

Equalizers

In general, the design of audio equalizers exists within the analog domain, and control over these devices is often manual. However, certain manufacturers of both digital and analog equalizers have begun to implement MIDI and digital control circuitry as a means of selecting user-programmed EQ curves from a device's front panel. These are automated using MIDI program messages from a sequencer or edited directly via SysEx or controller messages.

One example of a MIDI-controlled equalizer is the Rane MPE-28 from Rane Corporation (Fig. 8-5). This device is a single-channel, 28-band, 1/3-octave, MIDI programmable equalizer designed for MIDI performance systems. This digitally controlled analog equalizer interfaces to any MIDI controller or sequencer, allowing not only program changes, but also curve modifications to be made in real time using MIDI SysEx messages. The MPE-28 allows the user to store up to 128 EQ curves, which may be recalled either from the front panel controls, MIDI commands, another MPE 28, or from a computer. It is also capable of responding to expression controllers, such as MIDI continuous controllers and after touch.

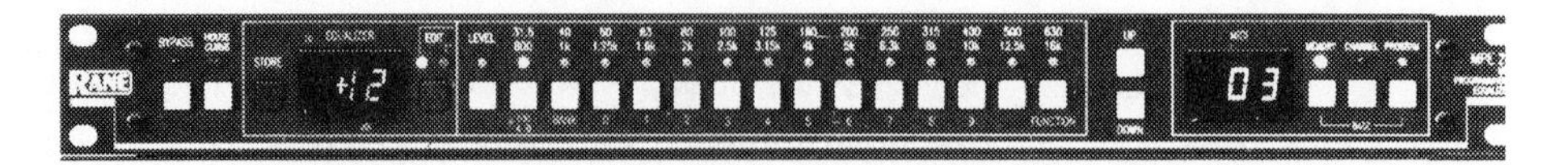

Fig. 8-5. The Rane MPE 28 MIDI programmable equalizer
(Courtesy of Rane Corporation)

The use of digital equalization can commonly be found as an integral function within newer multi-effects processors, digital musical instruments, and digital audio workstations. Stand-alone digital equalizer systems, such as the Yamaha DEQ7 dual-channel digital equalizer/filter-system (Fig. 8-6), are also available. The Yamaha DEQ7 is a digital equalizer which offers up to 30 preset filter configurations, including full graphic EQ, parametric EQ, shelving, notch, and dynamic-sweep filtering at a sampling rate of 44.1 kHz, with 32-bit internal processing circuitry. Up to 60 user-programmable EQ curves may be placed into its memory locations, with program recall and bulk dump capability being accessible via MIDI. The DEQ7's Digital I/O port also permits converterless operation for systems using the Yamaha digital audio format (such as the DMP-7D digital mixer).

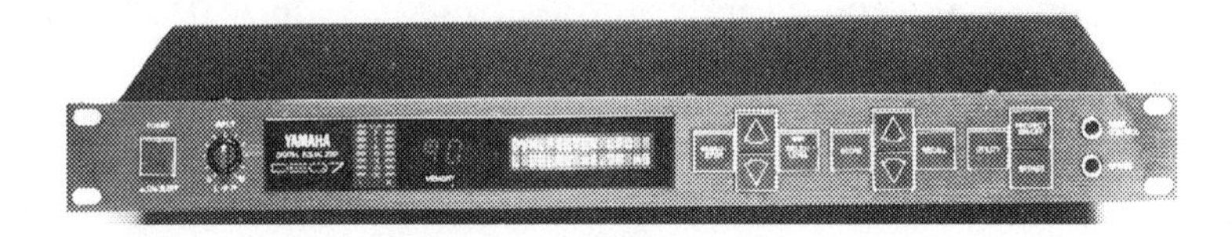

Fig. 8-6. The Yamaha DEQ7 digital equalizer. *(Courtesy of Yamaha Corporation)*

Dynamic Range Changers

As with current equalization circuitry, most MIDI-controlled dynamic range changers (compressors, limiters, and expanders) are found as a function within newer multi-effects processors, sample editors, and digital audio workstations. As of the date of writing, most analog-dynamic range devices do not incorporate MIDI preset locations, much less dynamic MIDI control over programming functions.

One dynamic range device that supports full MIDI implementation is the Drawmer M500 dynamics processor (Fig. 8-7). This multifunction signal processor is capable of performing a wide range of dynamic signal-control functions, including compression, limiting, expansion, gating, de-essing (frequency dependent limiting), autopanning, and fading. It also includes two filter sections which can be assigned to either the de-esser, gate, or split-band compressor programs.

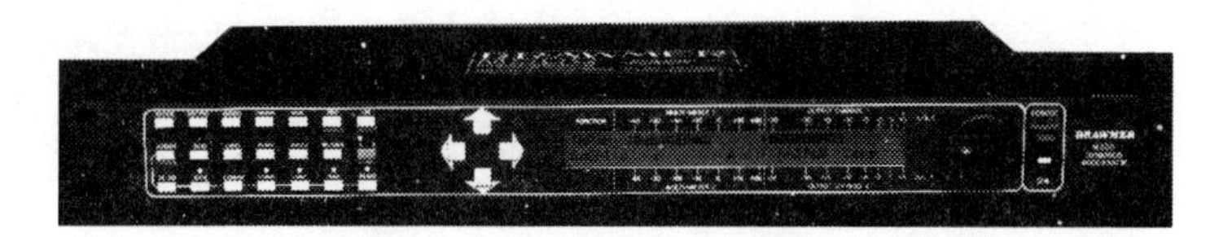

Fig. 8-7. The Drawmer M500 multifunction signal processor.
(*Courtesy of Quest Marketing*)

The M500 allows different functions to be simultaneously implemented in any logical sequence. It is fully programmable, offering the user 78 factory-preset signal processing patches, in addition to providing 50 user-programmable memory locations. Furthermore, this device offers MIDI implementation that includes recall of any memory preset via program-change messages, MIDI channel and receive mode assignments, control over processing parameters from a remote MIDI controller, selection of high/low-key split points, SysEx transmission/reception, and the setting of master or slave mode (allowing the unit to control or be controlled by another M500 device).

Effects Devices

By far the most commonly found MIDI controlled and programmable devices are digital effects processors. These devices are commonly used to either augment or modify a signal to improve its production value within a mix. Such a processor operates by converting analog signals into corresponding digital data. This data is then processed through user-specified algorithms (computer program for performing complex calculations upon digital data according to a predetermined pattern). The various parameters of an algorithm can be varied by the user to customize an effect to best suit a sound and/or music mix. Many of these effects devices are capable of processing a digital audio signal according to more than one algorithm, allowing the processor to produce a layering of multiple simultaneous effects.

One such processor is the Alesis MIDIVERB III (Fig. 8-8). This programmable 16-bit stereo digital effects device is capable of generating up to four simultaneous effects with an overall signal bandwidth of 15 kHz. These effects include 20 different reverb algorithms, up to 490 ms of delay in 1-ms increments, 6 kinds of chorus and 6 kinds of flanging, and an EQ section which includes a 6 dB per octave low-pass filter. The equalization is designed to roll off the high-frequency response of either the input or effect to simulate the high-frequency roll-off that occurs naturally in different acoustic environments.

Fig. 8-8. The MIDIVERB III simultaneous multi-effects processor. *(Courtesy of Alesis Studio Electronics)*

This device offers up to 200 preset effects registers made up of 100 nonerasable factory programs and 100 user-programmable presets. These effects may be mapped to a MIDI program-change number. Additionally, many of MIDIVERB III's parameters may be accessed in real time via MIDI SysEx.

The ART SGE MACH II (Fig. 8-9) is a stereo digital signal processor that offers up to 12 simultaneous audio functions with 20-bit internal processing. This device provides the user access to over 70 effects which can be stored within up to 200 memory locations. Some of these effects include: harmonic exciter, programmable equalizer, compressor & limiter, noise gate & expander, envelope filter, distortion & overdrive, reverb, delay, sampling (up to 2 seconds), pitch transposition, stereo panner, chorusing, and flanging.

Fig. 8-9. The ART SGE MACH II multi-effects processor. *(Courtesy of Applied Research & Technology, Inc.)*

Up to eight of the MACH II's control parameters can be accessed via MIDI controller messages (within up to eight simultaneous processing functions). In addition, a MIDI data monitoring mode allows the user to read real-time MIDI data messages (i.e., patch change, velocity, after touch, and continuous-controller messages) for any device connected to the MACH II's MIDI input.

A popular effects device that is designed for both electronic music and professional recording is the Lexicon PCM-70 software-based processor (Fig. 8-10). This device allows additional programs to be added as they become available through the installation of factory ROM's. Currently the device is shipped with more than 40 factory-generated programs, including concert hall, rich chamber, rich plate, infinite reverb, resonant chord, multiband delay, chorus, and echo effects. Using these factory programs as a basis, or by making use of available software-editing programs, up to 50 additional programs can be stored into different registers. Each program effect may be assigned to up to six voices, allowing the user to layer an effect (such as multiple choruses, echos, or delays).

Fig. 8-10. The Lexicon PCM-70 multiple effects processor. (*Courtesy of Lexicon, Inc.*)

The PCM-70 incorporates the innovative use of Dynamic MIDI into its operating structure. This allows as many as 10 effects or reverb parameters to be assigned to any MIDI controller or keyboard device, with partial parameter control being assignable by key velocity, pressure, or after touch. More than 80 different parameter types (such as delay times, beats-per-minute, feedback, wet-dry mix, high and low pass filters, room size, etc.) can be instantly accessed via MIDI in real time or from a sequencer.

Patches may be assigned and addressed from the front panel by using pad and *soft knob* controls. They can also be addressed via MIDI. In addition, this system incorporates a corresponding register table which allows any of the 128 MIDI-specified presets to be utilized for recalling a defined program.

Musical Instrument Processors

A range of effects devices are being designed expressly for use with musical instruments (such as keyboards and electric guitar systems). These devices offer a wide range of effects and multi-effects which can be accessed remotely via MIDI program changes or edited directly through the use of MIDI controller messages.

The Roland GP-16 Guitar Effects Processor (Fig. 8-11) is a digital, programmable multi-effects device that offers 16 individual effects. Up to 12 of these effects can be simultaneously processed. Effects routing can be programmed in any order within each effect block and stored independently, along with on/off and parameter settings within any of 128 patch memory locations.

Fig. 8-11. The Roland GP-16 Guitar Effects Processor.
(*Courtesy of Roland Corporation US*)

The GP-16 features two sets of unbalanced stereo 1/4-inch output jacks plus a balanced set of XLR jacks. The 1/4-inch jacks may be individually or simultaneously assigned to a patch, enabling the user to switch amplifier configurations during a performance. The balanced XLR connectors allow for the main output pair to be routed directly to a music-recording console.

By connecting the optional FC-100 MKII foot controller, any patch can be instantly accessed and selected. This controller can also be used to bypass or mute the processor entirely. Optional expression pedals may be connected to this controller, allowing the user to control any effects patch parameter in real time (such as chorus depth, picking filter cutoff frequency for a "whah" effect, or pitch shift for a pitch-bender effect).

Equipped with MIDI in, out, and thru jacks, the GP-16 is capable of accepting MIDI control change and program information through a foot controller and sequencer. Using MIDI mapping, each patch can be assigned to receive and transmit different program change numbers. MIDI bulk dump, load, and verify functions are also provided to allow the transfer of the GP-16's memory data to a sequencer via SysEx messages.

The Yamaha FX500 is another example of a digital multi-effects processor (Fig. 8-12). This device offers programmable compression, distortion, equalization, modulation, and reverb/delay effects stages. The compression, distortion, and EQ stages are basically single-function processors, while the modulation and reverb stages can be chained together to create up to five simultaneous effects within a single custom patch.

Fig. 8-12. Yamaha FX500 simul-effect processor (*Courtesy of the Yamaha Corporation*)

The FX500 contains a total of 91 memory locations, out of which locations 1–60 contain a range of factory-programmed effects. Presets 61–90 are reserved for user-programmed effects patches. Location 0 contains initial data used as a basis for creating an effects program from scratch.

Any of the memory locations can be accessed via MIDI program-change messages. It is also possible to directly control up to two different effects parameters in real time via two MIDI controllers (mod wheel, data fader, software-controller messages, etc.).

Summary

With the acceptance of MIDI in the music production industry, it has become common for digital effects and signal processing devices to offer a broad sound palette under the control of MIDI. As a tribute to this control medium, automation of these devices can be accomplished within any MIDI production studio or onstage environment from a sequencer or MIDI data controller. Should your system expand, the use of MIDI-equipped effects devices will often be flexible enough to accommodate your expanding production needs.

Chapter 9

Synchronization

In recent times, electronic music has evolved into being accepted as a standard production tool within the professional audio, video, and film media. For example, video postproduction houses routinely operate audio transports, video transports, and electronic musical instruments in computer-controlled tandem (Fig. 9-1), allowing the operator to create and refine a video soundtrack to perfection. One of the central concepts behind electronic music is the ability for MIDI-equipped sequencers, instruments, controllers, and effects devices to directly communicate with each other. Such MIDI-based sytems are also often required to directly interface with external real-time devices (such as multitrack tape machines, digital audio workstations, and VCRs). The method which allows both multiple devices and multiple media to maintain a dependable time relationship is known as *synchronization* (often abbreviated as *sync*).

Synchronization is the occurrence of two or more events at precisely the same time. With respect to analog audio and video systems, it is achieved by interlocking the transport speeds of two or more machines. Within computer-related systems (such as digital audio and MIDI), internal or external sync is commonly maintained through the use of a clock-timing pulse that is directly imbedded within the system's digital word structure (e.g., AES/EBU, MIDI sample dump, and MIDI). As there is commonly a need for maintaining synchronization between mechanical and digital devices, modern music and audio production often have resulted in the development of some rather ingenious forms of communication and data translation.

Within this chapter, we shall look into the various forms of synchronization for both analog devices and digital devices, in addition to current methods for maintaining sync between the two media.

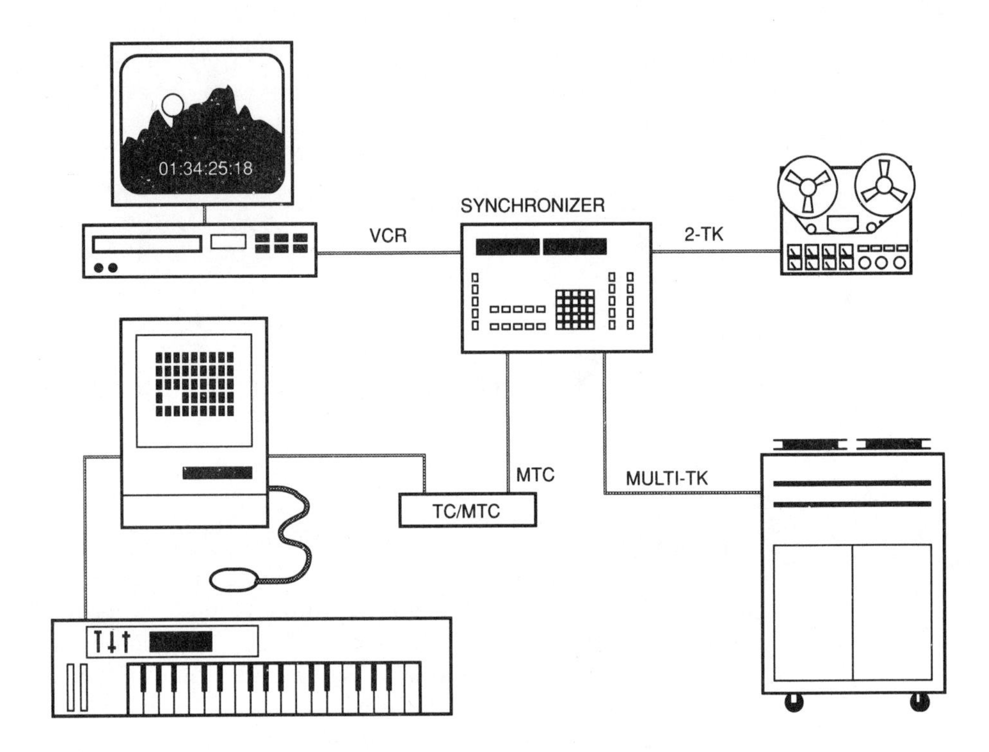

Fig. 9-1. Example of an integrated audio production system.

Synchronization between Analog Transports

Maintaining relative synchronization between analog tape transports does not require that all the transport speeds involved in the process be constant. However, it does require that they maintain the same relative speed.

Analog tape devices (unlike most of their digital counterparts) are unable to maintain a perfectly constant reproduction speed. For this reason, synchronization between two or more machines without some form of synchronous locking would be impossible over any reasonable program length. Synchronization would soon be lost as a result of such factors as voltage fluctuations and tape slippage. Thus, if production is to utilize multiple media, a means of interlocking these devices in synchronous time is essential.

SMPTE Time Code

The standard method of interlocking audio and video transports makes use of *SMPTE (Society of Motion Picture and Television Engineers) time code*. The use of time code allows for the identification of an exact location on a magnetic tape by assigning a digital address to each specified length over time. This identifying address code cannot slip and always retains its original-location identity, allowing for the continual monitoring of tape position to an accuracy of 1/30th of a second. These specified tape segments are called *frames* (a term taken from film production). Each audio or video frame is tagged with a unique identifying number, known as a *time-code address*. This eight digit address is displayed in the form 00:00:00:00, where the successive pair of digits represent Hours:Minutes:Seconds:Frames (Fig. 9-2). Additional information, which shall be discussed later within this chapter, can also be encoded within this address.

00:08:50:29			
HOURS	MINUTES	SECONDS	FRAMES

Fig. 9-2. Readout of a SMPTE time-code address.

The recorded time-code address is used to locate a position on magnetic tape in much the same fashion as a postal carrier makes use of an address for delivering the mail. For example, if a mail carrier is to deliver a letter addressed to our friend, Reggie, at 33 Music City Road, he or she knows precisely where to deliver the letter because of the assigned address numbers (Fig. 9-3A). Similarly, a time-code address can be used to locate specific locations on a magnetic tape. For example, let's assume that we would like to lay down the sound of a squealing car at a time-code address of 00:08:50:29 onto a time-encoded multitrack tape that begins at 00:01:00:00 and ends at 00:15:19:03 (Fig. 9-3B). Through the monitoring of the address code (in a fast shuttle mode), we can locate the position that corresponds to this tape address and perform the necessary function,"SCREEEE!".

Time-Code Word

The total of all time-encoded information recorded within each audio or video frame is known as a *time-code word*. Each word is divided into 80 equal segments called *bits*, which are numbered consecutively from 0 to 79. One word occupies an entire audio or video frame, so that for every frame there is a corresponding time-code address. Address information is contained in the digital word as a series of bits made up of binary ones and zeros. These bits are electronically generated as fluctuations (or shifts), in the voltage level of the time code's data signal. This method of encoding serial information is known as *biphase modulation* (Fig. 9-4). When the recorded biphase signal shifts either up or down at the extreme edges of a clock period, the pulse is coded as a binary 0. A binary 1 is coded for a bit whose signal pulse shifts halfway through a clock period. A

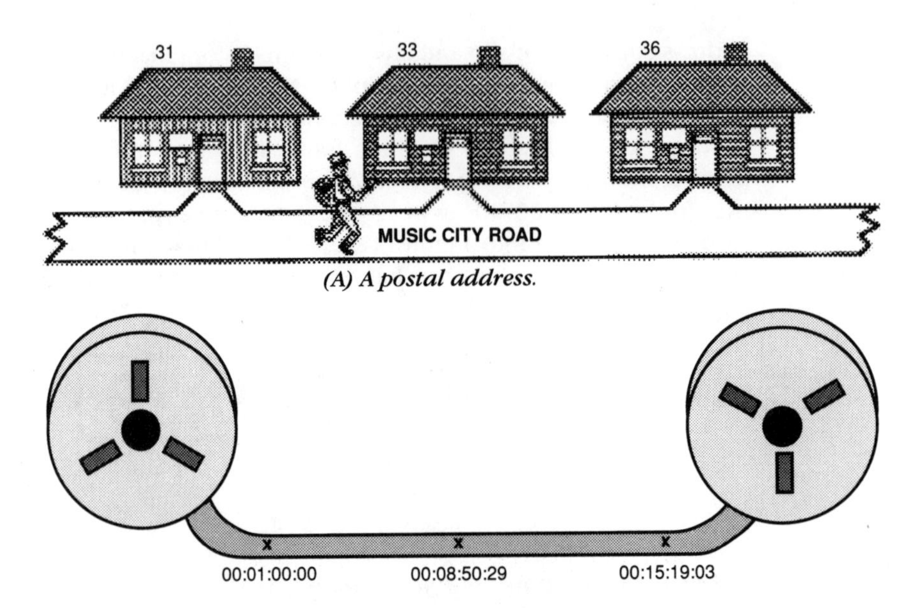

(A) A postal address.

(B) A SMPTE time-code address on magnetic tape.

Fig. 9-3. Relative address locations.

positive feature of this method of encoding is that detection relies on shifts within the pulse and not on the pulse's polarity. This means that time code can be read in either the forward or reverse direction, and at fast or slow shuttle speeds.

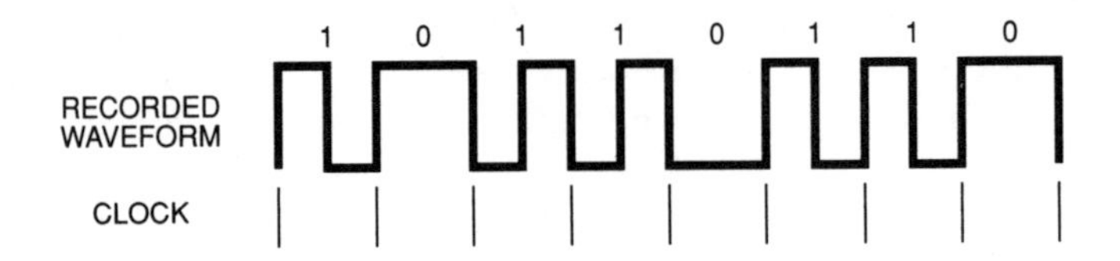

Fig. 9-4. Biphase modulation encoding.

The 80-bit time-code word (Fig. 9-5) is further subdivided into groups of four bits, with each group representing a specific coded piece of information. Within each of these 4-bit segments is the encoded representation of a decimal number ranging from 0 to 9, written in binary-coded decimal (BCD) notation. When a time-code reader detects the pattern of ones and zeros within a 4-bit group, it interprets the information as a single decimal number. Eight of these 4-bit groupings combine to constitute an address in hours, minutes, seconds, and frames.

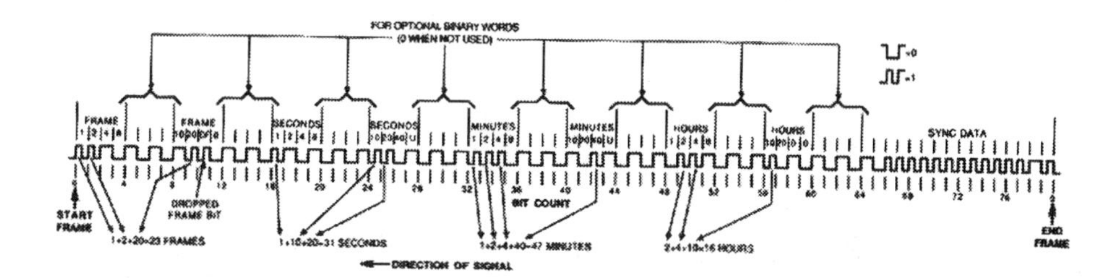

Fig. 9-5. Biphase representation of the SMPTE/EBU time-code word.

The 26 digital bits which make up the time-code address are joined by an additional 32 bits called ***user bits***. This additional set of encoded information, which is also represented in the form of an 8-digit number, has been set aside for time-code users to enter personal ID information. The SMPTE Standards Committee has placed no restrictions on the use of this slate code, which can contain such information as date of shooting, take ID, reel number, etc.

Another form of information encoded within the time-code word is *sync data*. The sync data is found in the final 16 bits of a time-code word, and is used to define the end of each frame. Because time code can be read in either direction, sync data also signals the controlling device about which direction the tape or device is moving.

Time-Code Frame Standards

In productions using time code, it is important that the readout display be directly related to the actual elapsed program time, particularly when dealing with the exacting time requirements of broadcasting. The black-and-white (monochrome) video signal operates at a frame rate of exactly 30 frames per second (fr/second). This monochrome rate is known as nondrop code. If this time code is read, the time-code display, program length, and actual clock on the wall will all be in agreement.

However, this simplicity was broken when the National Television Standards Committee set the frame rate for the color video signal at approximately 29.97 fr/second. This would mean that when a time-code reader set to read the monochrome rate of 30 fr/second is used to read a 29.97-fr/second color program, the time-code readout would pick up an extra .03 frames for every passing second (30 – 29.97 = .03 fr/second). Over the duration of an hour, the address readout will differ from the actual clock by a total of 108 frames or 3.6 seconds.

To correct this discrepancy and regain an agreement between the time-code readout and the actual elapsed time, a means of frame adjustment must be introduced into the code. Since the object is to drop 108 frames over the course of an hour, the code used for color has come to be known as *drop-frame code*.

In correcting this timing error, two frame counts for every minute of operation are omitted, with the exception of minutes 00, 10, 20, 30, 40, and 50. This has the effect of adjusting the frame count to agree with the actual elapsed program duration.

In addition to the color 29.97 drop-frame code, a 29.97 nondrop frame color standard is also commonly used within video production. When using this *nondrop time code*, the frame count will always advance one count per frame, with no drops in count. This will result in a disagreement between the frame count and the actual clock-on-the-wall time over the course of the program. However, this method has the distinct advantage of easing the time calculations often required within the video-editing process because no frame compensations need to be taken into account for dropped frames.

Another frame-rate format used throughout Europe is *EBU (European Broadcast Union) time code*. EBU utilizes the SMPTE 80-bit code word. However, it differs in that it uses a 25-fr/second frame rate. Since both monochrome and color video EBU signals run at exactly 25 fr/second, there is no necessity for an EBU drop-frame code.

The film media makes use of a standardized 24-fr/second format that differs from SMPTE time code. Many newer synchronization and digital audio devices offer film sync and calculation capabilities.

LTC and VITC Time Code

There are currently two major methods for encoding time code onto magnetic tape for broadcast and production use: LTC and VITC.

Time code which is recorded onto an audio or video cue track is known as *longitudinal time code (LTC)*, commonly misnomered in the electronic music, recording, and video media as simply "SMPTE." LTC encodes the biphase time-code signal onto an analog audio or cue track as a modulated square-wave signal with a bit rate of 2400 bits/second.

The recording of a perfect square wave onto a magnetic audio track is, under the best of conditions, difficult. For this reason the SMPTE has set forth a standard allowable risetime of 25 ± 5 microseconds for the recording and reproduction of code. This is equivalent to a signal bandwidth of 15 kHz, well within the range of most professional audio-recording devices.

Variable speed time-code readers are commonly able to decode time-code information at shuttle rates ranging from 1/10th to 100 times normal playspeed. This is effective for most audio applications. However, within video postproduction, it is often necessary to monitor a video tape at slow or still speeds. Because LTC cannot be read at speeds slower than 1/10th to 1/20th normal playspeed, two methods can be used for reading time code. The time-code address may be burned into the video image of a copy worktape, where a character generator is used to superimpose the corresponding address within an on-screen window (Fig. 9-6). This window dub allows time code to be easily identified, even at very slow or still picture shuttle speeds.

Fig. 9-6. Video image showing burned-in time-code window.

Another method used by major production houses is to stripe the picture with *VITC (Vertical Interval Time Code)*. VITC makes use of the same SMPTE address and user-code structure as does LTC. However, it is encoded onto video tape in an entirely different signal form. This method actually encodes the time-code information within the video signal itself inside a field located outside of the visible-picture scan area, known as the ***vertical blanking interval***. Since the time-code information is encoded within the video signal itself, it is possible for 3/4-inch or 1-inch helical-scan video recorders to read time code at slower speeds and still frame. Because the time code can be accurately read at all speeds, this added convenience opens up an additional track on a video recorder for audio or cue information and eliminates the need for a window dub.

In most situations LTC code is preferred for audio electronic music production and standard video production since it is a more accessible and cost-effective protocol.

Jam Sync/Restriping Time Code

As we have seen, LTC operates by recording a series of square-wave pulses onto magnetic tape. Unfortunately, it is not completely a simple matter to record the square waves without moderate to severe waveform distortion.

Although time-code readers are designed to be relatively tolerant of waveform amplitude fluctuations, the situation is severely compounded when the code is dubbed (copied) by one or more generations. For this reason a special feature, known as *jam sync*, has been incorporated into most time-code synchronizers.

It is basically the function of the jam-sync process to regenerate fresh code to identically match old time-code address numbers during the dubbing stage or reconstruct defective sections of code (Fig. 9-7).

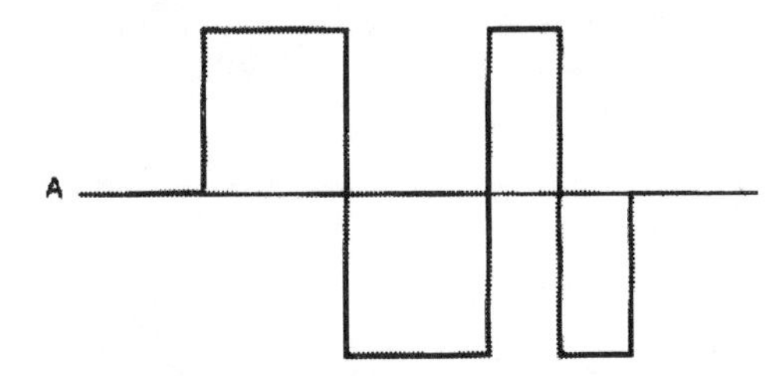

(A) Reproduction of original biphase signal.

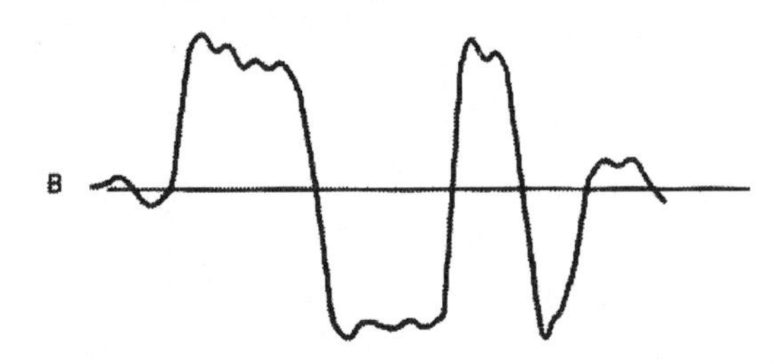

(B) Reconstructed jam-sync signal.

Fig. 9-7. Representation of the recorded biphase signal.

Currently, there are two common forms of jam sync in use: one-time jam sync and continuous jam sync.

One-time jam sync refers to a mode in which, upon receiving valid time-code, the generator's output is initialized to the first valid address number received and begins to count in an ascending order on its own in a freewheeling fashion. Any deteriorations or discontinuities in code are ignored, and the generator will produce fresh uninterrupted address numbers.

Continuous jam sync is used in cases where it is required that the original address numbers remain intact and not regenerated into a contiguous address count. Once the reader is activated, the generator will update the address count and each frame in accordance with the incoming address.

Synchronization Using Time Code

To achieve a frame-for-frame time-code lock among multiple audio, video, and film transports, it is necessary to employ a device known as a *synchronizer* (Fig. 9-8). The basic function of a synchronizer is to control one or more tape or film transports (designated as *slave machines*), whose position and tape speed are made to accurately follow one specific transport (designated as *the master*).

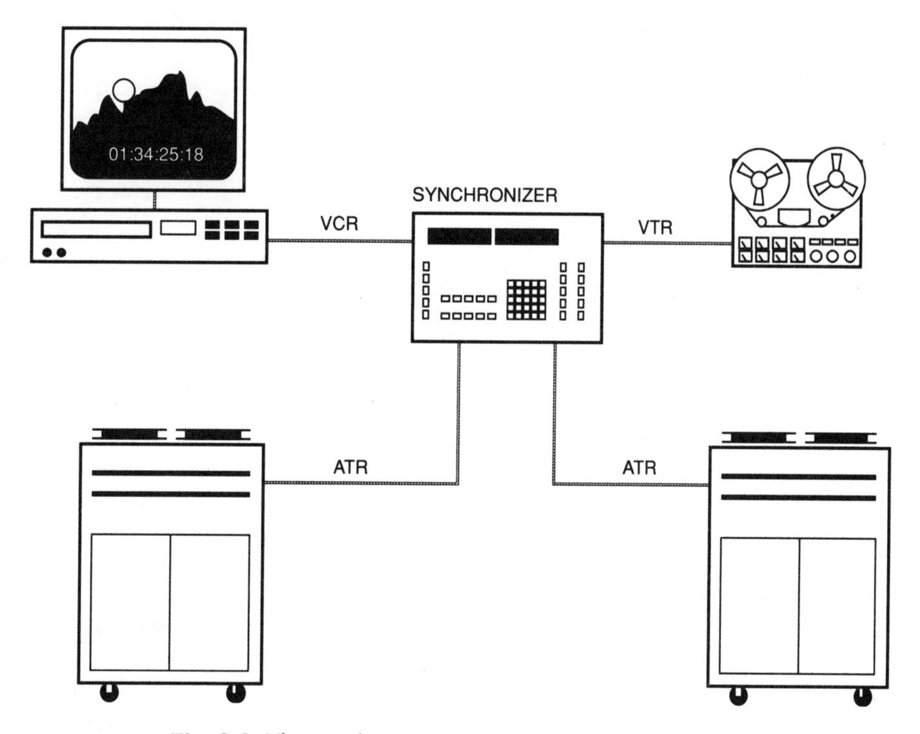

Fig. 9-8. The synchronizer within time-code production.

The most common synchronizer types found within modern postproduction are those of the chase synchronizer, the control synchronizer, and the *edit decision list* (*EDL*) synchronizer.

The *chase system* of synchronization requires that a slave transport chase the master under all conditions. This system is a bit more sophisticated than a play-speed-only system since the synchronizer has additional control over the slave transport's operating functions, and is often able to read the system time code for locating areas in the fast-shuttle/wind mode. This enhancement allows the slave transport to automatically switch from the play mode back into the cue (search) mode to chase the master under all conditions and resync to it when the master is again placed into the play mode.

The *control synchronizer* (Fig. 9-9), as a device, places an emphasis upon control over all transport functions found within a synchronized system. Operating from a central keyboard, the control synchronizer provides such control options as:

- *Machine selection*: Allows the selection of machine(s) to be involved in the synchronization process, as well as allowing for the selection of a designated master.
- *Transport control*: Allows for conventional remote control of functions over any or all machines in the system.
- *Locate*: A transport command which causes all selected machines to locate automatically to a selected time-code address.
- *Looping*: Enters a continuous repeat cycle (play, repeat, play) between any two address cue points which are stored into memory.
- *Offset*: Permits the correction of any difference in time code which exists between program material (i.e., to adjust relative frame rates by ± X frames to achieve or improve sync).
- *Event points*: A series of time-code cue points which are entered into memory for use in triggering a series of function commands (i.e., start slave or mastering machine, record in/out, insert effects device, or any externally triggered device).
- *Record in/out*: An event function allowing the synchronizer to take control over transport record/edit functions, enabling tight record-in/out points to be repeated with frame accuracy.

Fig. 9-9. The Adams-Smith Zeta-Three audio/video/MIDI synchronizer with the Zeta-Remote autolocator/controller. (*Courtesy of Adams-Smith*)

The most recent entry into the field of audio synchronization is the EDL controller synchronizer. This development has evolved out of the use of the EDL within the on-line video-editing process, and is most commonly found within the video and audio-for-video postproduction suite.

The EDL is based upon a series of edit commands which are entered and stored within computer memory as a database. Once the information has been entered, the system will electronically control, synchronize, and switch all of the associated studio transports, exert control over edit-in/out points, tape positions, time-code off set instructions, etc., with complete time-code accuracy and repeatability.

Set Up for Production Using Time Code

Generally, in audio production, the only required connection between the master machine and the synchronizer is a time-code signal feed (Fig. 9-10). If a control synchronizer is used, a number of connections are required between the slave transports and synchronizer. These include provisions for the time-code signal, full logic-transport control, and a DC control-voltage signal (for driving a slave machine's servo capstan).

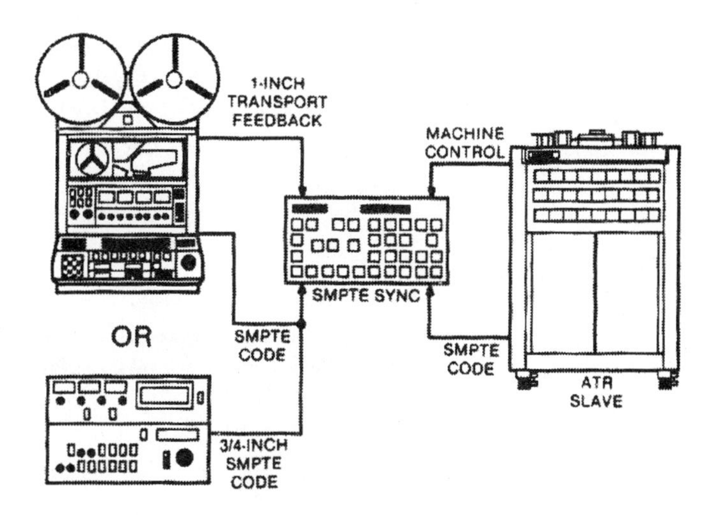

Fig. 9-10. System interconnections for synchronous audio production.

Distribution of LTC Signal Lines

Longitudinal SMPTE time code may be distributed throughout the production and postproduction system in a similar fashion to any other audio signal. It can be routed through any audio path or patch point via normal two-conductor shielded cables; and, since the time-code signal is biphase or symmetrical in nature, it is immune to problems of polarity.

Time-Code Levels

One problem that can potentially plague systems using time code is cross talk, which arises from high-level time-code signals interfering with adjacent audio signals or recorded tape tracks. Currently, no industry standard levels exist for the recording of time code onto magnetic tape. However, the following levels have proven over time to give the best results in most cases.

Table 9-1. Optimum Time-Code Recording Levels

Tape Format	Track Format	Optimum Rec. Level
ATR	Edge track (highest number)	–5 VU to –10 VU
3/4-inch VTR	Audio 1 (L) track or time code track	–5 VU to 0 VU
1-inch VTR	Cue track or audio 3	–5 VU to –10 VU

(Note: Do not record SMPTE time code onto a track being noise reduced, such as Dolby or dbx. Bypass the noise-reduction circuitry on the time-code track, or if necessary, manually patch around the circuit. If the VTR being used is equipped with AGC [Automatic Gain Compensation], override the AGC and adjust the signal gain controls manually.)

Non-MIDI Synchronization

There are several types of synchronization used by older electronic instruments and devices that were designed before the MIDI standard was implemented.

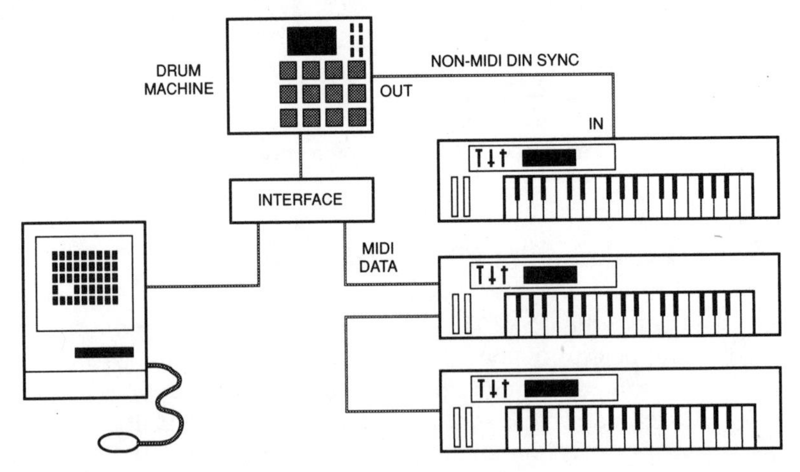

Fig. 9-11. One of the many possible methods for syncing MIDI and non-MIDI devices.

Although sync between these non-MIDI and MIDI instruments (Fig. 9-11) can be a source of mild to major aggravation, many of these older devices remain within MIDI setups because of their unique and wonderful sounds. Such older sync standards are still in wide use within both MIDI and non-MIDI systems to allow humans to easily determine tempo.

Click Sync

Click sync or *click track* refers to the metronomic audio clicks that are generated by electronic devices to communicate tempo. These are produced once per beat or once per several beats (as occurs in cut time or compound meters).

Often, a click or metronome is designed into a MIDI interface which produces an audible tone (or series of tone pitches) that can be heard and used as a tempo guide. Such interfaces and other sync devices offer an unbalanced clicks-sync out jack that can be fed to a mixer and blended into a speaker or headphone monitor mix. This allows one or more musicians to keep in tempo to a sequenced composition. This click can also be recorded onto an unused track of a multitrack recorder for use within a project that involves both sequenced and live music.

Certain sync boxes and older drum machines are capable of syncing a sequence to a live or recorded click track. Such devices are able to determine the tempo based upon the frequency of the clicks and will output a MIDI start message (once a sufficient number of click pulses have been received for tempo calculation). A MIDI stop message may be transmitted by such a device whenever more than two clicks have been missed at or below the slowest possible tempo (30 BPM).

This sync method does not work well with rapid tempo changes. Chase resolution is limited to one click per beat (1/24th the resolution of MIDI clock). Thus, it is best to use a click source relatively constant in tempo.

TTL and DIN Sync

One of the most common methods of synchronization between early sequencers, drum machines, and instruments before the adoption of MIDI was TTL 5-volt sync. This method makes use of 5-volt clock pulses in which a swing from 0 to 5 volts represents one clock pulse.

Within this system, a musical beat is divided into a specific number of clock pulses per quarter note (i.e., 24, 48, 96, and 384 PPQN) which varies from device to device. For example, *DIN sync* (a form of TTL sync), which is named after the now famous 5-pin DIN connector, transmitted at 24 PPQN.

TTL may be transmitted in either one of two ways. The first and simplest method makes use of a single conductor through which a 5-volt clock signal is sent. Quite simply, once the clock pulses are received by a slave device, it will start and synchronize to the incoming clock rate. Should these pulses stop, the devices will also stop and wait for the clock to again resume. The second method makes use of two conductors both of which transmit 5-volt transitions. However, within the system, one line is used to constantly transmit timing information, while the other is used for start/stop information.

MIDI-Based Synchronization

Within current MIDI production, the most commonly found form of synchronization makes use of the MIDI protocol for the transmission of sync messages. These messages are transmitted along with other MIDI data over standard MIDI cables, with no need for additional or special connections.

MIDI Sync

The system that is most commonly used within a basic MIDI setup is known as *MIDI Sync*. This sync protocol is primarily used for locking together the precise timing elements of MIDI devices within an electronic music system and operates through the transmission of real-time MIDI messages over standard MIDI cables.

As with all forms of synchronization, one MIDI device must be designated to be the master device to provide the timing information to which all other slaved devices are locked.

MIDI Real-Time Messages

MIDI real-time messages are made up of four basic types, which are each one byte in length: timing clock, start, continue, and stop messages.

The *timing-clock message* is transmitted to all devices within the MIDI system at a rate of 24 PPQN.

Recently, a few manufacturers have begun to develop devices which generate and respond to 24 clock signals per metronomic beat. This method is used to improve the system's timing resolution and simplify timing when working in nonstandard meters (i.e., 3/8, 5/16, 5/32, etc.).

The *MIDI start command* instructs all connected devices to start playing from the beginning of their internal sequences upon the receipt of a timing-clock message. Should a program be in mid-sequence, the start command will reposition the sequence back to its beginning, at which point it will begin to play.

Upon the transmission of a *MIDI stop command* all devices within the system will stop at their current position and wait for a message to follow.

Following the receipt of a MIDI stop command, a *MIDI continue message* will instruct all sequencers and/or drum machines to resume playing from the precise point at which the sequence was stopped.

Certain older MIDI devices (most notably drum machines) are not capable of sending or responding to continue commands. In such cases, the user must either restart the sequence from its beginning or manually position the device to the correct measure.

Song Position Pointer

In addition to MIDI real-time messages, the *song position pointer* (*SPP*) is a MIDI system common message, which acts as a relative measure of musical time (as expressed in measures) that has passed since the beginning of the sequence. The SPP is expressed as multiples of 6 timing-clock messages, and is equal to the value of a 16th note.

Song position pointer allows a compatible sequencer or drum machine to be synchronized to an external source from any position within a song of 1024 measures or less. Thus, when using SPP, it is possible for a sequencer to chase and lock to a multitrack tape from any specific measure within a song.

Within such a MIDI/tape setup (Fig. 9-12), a specialized sync tone is transmitted by a device that encodes the sequencer's SPP messages and timing data directly onto tape as a modulated signal. Unlike SMPTE time code, the means by which manufacturers encode this data onto tape is not standardized. This lack of a standard format could prevent SPP data written by one device from being decoded by another device that uses an incompatible proprietary sync format.

Many SMPTE-to-MIDI synchronizers (such as Adams-Smith's Zeta-3, J.L.Cooper's PPS-100, or Roland's SBX-80) can also be used to instruct a slaved sequencer, drum machine, or other device to locate to a specific position within the sequence (as defined by the number of 16th notes from the beginning of a song). Once the device(s) has located to the correct position, the system will stop and wait until a continue message and ensuing timing clocks have been received.

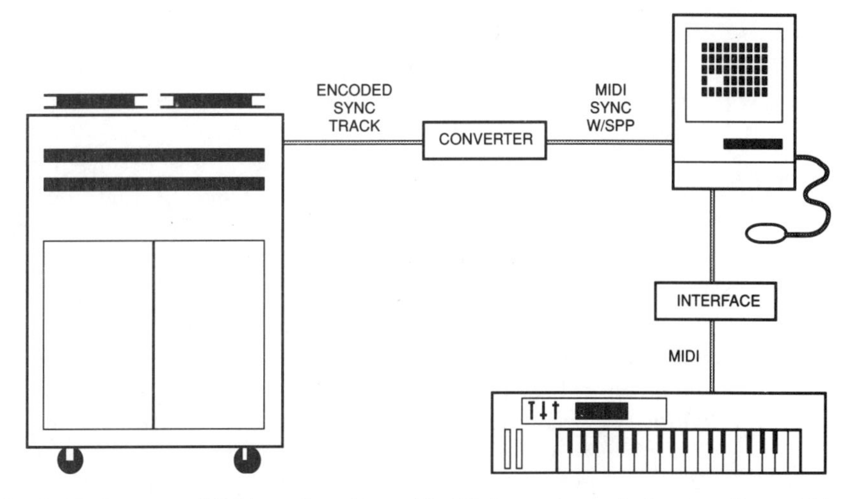

Fig. 9-12. A tape-to-MIDI synchronizer with SPP is used as a timing interface within studio production.

To vary tempo, while maintaining sync between the sequencer and SMPTE control track, many SMPTE-to-MIDI synchronizers can be preprogrammed to create a tempo map that provides for tempo changes at specific SMPTE times. Once the SPP control track is committed to tape, however, the tape and sequence are locked to this predetermined tempo or tempo map.

SPP messages are usually transmitted only while the MIDI system is in the stop mode, in advance of other timing and MIDI continue messages. This is due to the relatively short period of time that is needed for the slaved device to locate to the correct measure position.

Certain devices, such as earlier sequencers and drum machines, do not respond to SPPs. In order to take advantage of the sync benefits, it is best that these devices be slaved to a master timing device that will respond to these pointers.

FSK

In the pre-MIDI days of electronic music, musicians discovered that it was possible to sync instruments that were based upon such methods as TTL 5-volt sync to a multitrack tape recorder. This was done by recording a square-wave signal onto tape (Fig. 9-13A) which could serve as a master sync pulse. The most common pulse in use was 24 and 48 PPQN; therefore, the recorded square wave consisted of a 24- or 48-Hz signal.

Although this system worked, it was not without its difficulties. This is because the devices being synced to the pulse relied upon the integrity of the square wave's sharp transition edges to provide the clock. Tape is notoriously bad at reproducing a square wave (Fig. 9-13B). Thus, the poor frequency response and reduced reliability at low frequencies mandated that a better system for syncing MIDI to tape be found. The initial answer was in *frequency shift keying*, better known as *FSK*.

FSK works in much the same way as the TTL-sync track. However, instead of recording a low-frequency square wave onto tape, FSK makes use of two, high-frequency square-wave signals for marking clock transitions (Fig. 9-13C). In the case of the MPU-401/compatible interface, these two frequencies are 1.25 kHz. and 2.5 kHz. The rate at which these pitches alternate determines the master timing clock to which all slaved devices are synced. These devices are able to detect a change in modulation, convert these into a clock pulse, and advance its own clock accordingly.

As FSK makes use of a higher frequency range than its earlier counterpart, it is far more resistant to signal deformation and low-frequency roll off. However, when reproducing FSK from tape, the tape recorder's speed should be kept within reasonable tolerances because a wide variance may make it difficult for the slaved device to recognize the frequency shift as a valid clock transition. For this reason, tempo can only be varied within reasonable limits when varying the speed of a tape recorder.

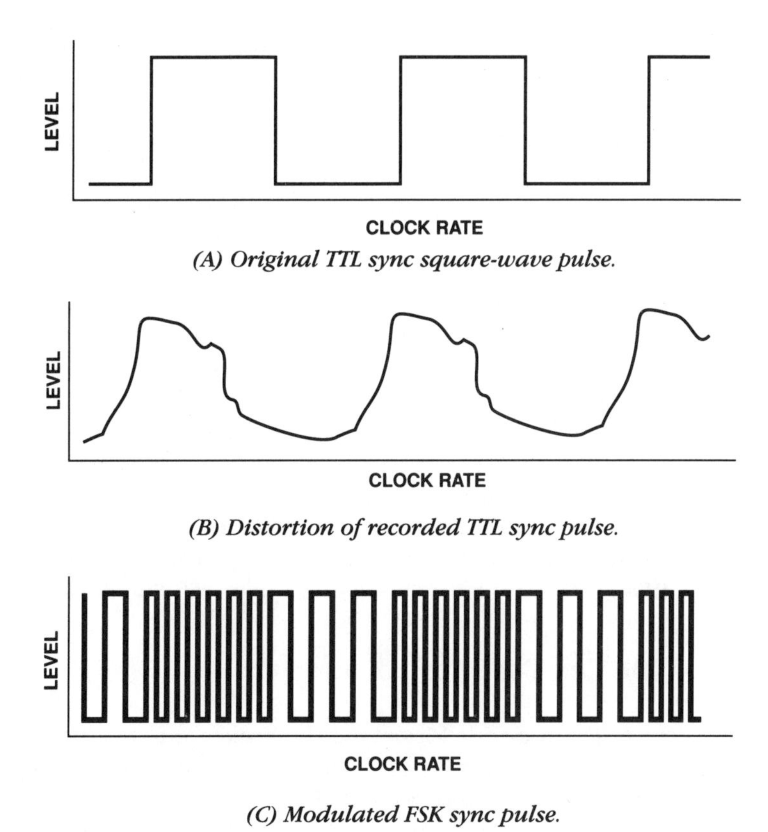

(A) Original TTL sync square-wave pulse.

(B) Distortion of recorded TTL sync pulse.

(C) Modulated FSK sync pulse.

Fig. 9-13. TTL and FSK sync track waveforms.

The level at which FSK can be recorded onto tape is also a consideration. Too low a level may make it difficult for the slaved device(s) to accurately decipher the signal. Signals that are too high in level may cause problems with distortion or crosstalk. The most commonly accepted VU levels are –3 dB for semiprofessional recorders and –10 dB for professional equipment, and if it can at all be avoided, do not record FSK onto a track containing noise reduction.

MIDI Time Code

For decades, the SMPTE time code has been the standard timing reference within audio and video production. This is because it is an absolute timing reference which remains constant throughout a program. On the other hand, both MIDI clock and SPP are relative timing references which vary with both tempo and tempo changes. As most studio-based operations are referenced to SMPTE time-code addresses, as opposed to the beats in a musical bar, it would be extremely tedious for a studio engineer or musician to convert between the two timing systems, when cueing or triggering a specific event.

In order for MIDI-based devices to operate on an absolute timing reference independent of tempo, *MIDI time code* or *MTC* was developed. Created by Chris Meyer and Evan Brooks of Digidesign, MIDI time code provides a cost-effective and easily implemented means for translating SMPTE time code into MIDI messages. It also allows for time-based code and commands to be distributed throughout the MIDI chain to those devices or instruments which are capable of understanding and executing MTC commands.

MTC does not replace, but rather is an extension of MIDI 1.0 since it makes use of existing message types that were either previously undefined or were being used for other, nonconflicting purposes such as the sample-dump standard. Most existing MIDI devices need not, and will never, directly make use of MTC. However, newer devices and time-related program packages, which read and write MTC, are currently being developed.

MTC makes use of a reasonably small percentage of the available MIDI bandwidth (about 7.68 percent at 30 fr/second). Although it is able to travel the same signal path as conventional MIDI data, it is recommended (if at all possible within a 32-channel system) that the MTC signal path be kept separate from the MIDI performance path to reduce the possibility of data overloading or delay.

MIDI Time-Code Control Structure

The MIDI time-code format may be broken into two parts: time code and MIDI cueing. The time-code capabilities of MTC are relatively straightforward and allow both MIDI and non-MIDI devices to attain synchronous lock or to be triggered via SMPTE time code. MIDI cueing is a format that informs MIDI devices of events to be performed at a specific time (such as load, play, stop, punch-in/out, and reset). This means of communication envisions the use of intelligent MIDI devices which are able to prepare for a specific event in advance, and then execute a command on cue.

MTC Sequencer

To control existing MIDI or triggerable devices which do not recognize MTC, an MTC sequencer must be used. Such a device or program package enables a time-encoded cue list to be compiled by the user and at specified SMPTE times, send related MIDI commands to the proper device within the system. Such MIDI messages might include note on, note off, song select, start, stop, program changes, etc. Non-MIDI equipment which can be triggered, such as CD players, mixing consoles, lighting, and audio tape recorders, may also be controlled.

One such software-based MTC sequencer is the Q-Sheet A/V automation system from Digidesign. Using appropriate MIDI equipment and a Macintosh computer (minimum of 512K with 800K of disk space), Q-Sheet A/V is able to automate many aspects of audio production, including effects units, keyboards, samplers, MIDI-based mixers, or any device that can be controlled via MIDI commands.

Q-Sheet A/V is much like a music sequencer since it records, stores, and plays back various types of MIDI events using individual tracks. Although most music sequencers combine multiple tracks to assemble a song, Q-Sheet A/V combines one or more tracks to create a cue list of MIDI events. An event might be placed into this list as the real-time movement of a MIDI-controlled fader, the setup of an effects parameter using SysEx messages, the adjustment of a delay setting or simply a note-on/off command for a sampler or synth. MIDI file-compatible sequences may also be loaded into the list and played with SMPTE-locked precision. Once the music or visual sound track composition has been created, the entire cue list can be used to trigger and automate a mix-under time-code command with quarter-frame accuracy at all SMPTE frame rates.

Intelligent MTC Peripheral

An intelligent MTC device would be capable of receiving an MTC-based cue list from a central controller and then trigger itself appropriately upon receiving a corresponding MTC address time. Additionally, such a device might be able to change its programming in response to the cue list, or prepare itself for ensuing events.

MIDI Time-Code Commands

MIDI time code makes use of three message types: quarter frame messages, full messages, and MIDI cueing messages.

Quarter-Frame Messages

Quarter-frame messages are transmitted only while the system is running in real or vary-speed time, and in either the forward or reverse directions. In addition to providing the system with its basic timing pulse, four frames are generated for every SMPTE time-code field. This means that should we decide to use drop-frame code (30 fr/second), the system would transmit 120 quarter-frame messages per second.

Quarter-frames messages should be thought of as groups of eight messages, which encode the SMPTE time in hours, minutes, seconds, and frames. Since eight quarter frames are needed for a complete time-code message, the complete SMPTE time is updated every two frames. Each quarter-frame message contains two bytes; the first being F1 (the quarter-frame common header), and the second byte contains a nibble (four bits) that represents the message number 0–7, and a nibble for each of the digits of a time field (hours, minutes, seconds, or frames).

Full Messages

Quarter-frame messages are not sent while in the fast-forward, rewind, or locate modes, because they would unnecessarily clog or outrun the MIDI data lines. When in any of these shuttle modes, a full message (which encodes the complete time-code address within a single message) is used.

Once a fast shuttle mode is entered, the system will generate a full message and then place itself into a pause mode until the time encoded device has arrived at its destination. After the device has resumed playing, MTC will again begin sending quarter-frame messages.

MIDI Cueing Messages

MIDI cueing messages are designed to address individual devices or programs within a system. These 13-bit messages may be used to compile a cue or edit decision list, which in turn instructs one or more devices to play, punch in, load, stop, etc. at a specific time. Each instruction within a cuing message contains a unique number, time, name, type, and space for additional information. At present only a small percentage of the possible 128 cueing-event types have been defined.

Direct Time Lock

Direct time lock, also known as *DTL*, is a synchronization standard that allows Mark of the Unicorn's Mac-based sequencer, Performer, to lock to SMPTE through a converter which supports this standard. The following is a detailed technical specification:

There are two messages associated with direct time lock: tape position and frame advance.

The *tape-position message* is transmitted when the time-code source (such as a tape machine) is started, whereafter the converter achieves lock. This message is implemented as a SysEx message and specifies the tape's SMPTE position in hours, minutes seconds, and frames (HH:MM:SS:FF).

<tape position> :==

<F0H> <28H> <15H> <HR> <MIN> <SEC> <FRM> <00> <F7H>

<HR> is one byte, specifying the hour (0–23)

<MIN> is one byte, specifying the minute (0–59)

<SEC> is one byte, specifying the seconds (0–59)

<FRM> is one byte, specifying the frame (0–29)

The tape-position message is a starting reference for the frame-advance messages. It need not be sent in sync with anything. A converter may wish to send the message a few frames before the specified frame is reached.

The *frame-advance message* is transmitted once each frame. The first frame advance sent after the tape-position message corresponds to the beginning of the frame specified within the tape-position message. Successive frame advances correspond to successive frames.

<frame advance> :== <F8H>

The frame-advance message is the same as the MIDI clock message used in normal MIDI sync. This is a real-time message as defined by MIDI, and may be inserted in the middle of a normal MIDI message for minimum timing delay. Optionally, the converter may send periodic tape messages between frame advance, while the tape is running. It is recommended by Mark of the Unicorn that periodic tape-position messages be sent every half second to a second.

Once a tape-position message has been received, the sequence will chase that point, and playback will be readied. Once frame messages are received, Performer will advance in sync with the master code. If more than 8 frames pass without a frame-advance message, Performer will assume that playback has stopped. It will then stop the sequence and wait for a new position message.

If another position message is received, Performer compares it to the position of its last position message. If they are the same, Performer will immediately continue playing. Otherwise Performer will chase to the desired sequence location. If the tape-position message is not close to the current location, Performer will immediately stop and chase to the new location, where it will ready playback and begin looking for frame-advance messages.

Problems may be encountered at slow tape motions (tape rocking, etc.), and it is not recommended that frame-advance messages be sent at high speeds (i.e., when the tape is rapidly cueing). At these times frame advances are stopped and a new tape-position message is sent once the tape has returned to normal playback speed.

Advanced Direct Time Lock

An advanced form of direct time lock (known as *DTLe*) has been incorporated into Performer 3.4 and higher versions. Enhanced direct time lock is used to synchronize Performer to the MIDI time piece MIDI interface, via SMPTE time code.

DTLe differs from standard direct time lock because it transmits four frame-advance messages per SMPTE frame instead of one. In addition, DTL's tape-position (full-frame) message has been expanded to include SMPTE frame count and an identifer for the device (within the MIDI time piece network) that transmits DTLe.

DTLe offers distinct advantages over its predecessor since it allows Performer 3.4 to establish synchronous lock with the tape machine without stopping the transport. Thus, upon pressing Performer's play button, the program will jump into sync while the audio or video machine is running. In addition, as the MIDI time piece transmits a tape-position message approximately once every second, the user is less likely to encounter problems, such as drop-outs or drift.

Performer Version 3.4 or higher is required to lock with a MIDI time piece via enhanced DTL. Although the time piece will not support old DTL, Performer 3.4 will support it, through the use of a standard DTL converter.

SMPTE to MIDI Converter

A SMPTE-to-MIDI converter serves to read either longitudinal or vertical SMPTE time code and convert it into such MIDI-based sync protocols as SPP, direct time lock, and MIDI time code. SMPTE-to-MIDI converters may be designed as a self-contained, stand-alone device, or within a SMPTE time-code synchronizer. In the future, it is envisioned that such converters may be directly integrated into such non-MIDI devices as audio or video tape recorders. These devices do not convert MIDI-based sync to SMPTE since this would be of little practical value within most electronic music setups.

One such converter, the PPS-1 (Poor Person's SMPTE) Version Three MIDI sync box from J.L. Cooper Electronics (Fig. 9-14) offers such features as MIDI sync-to-tape with SPP, SMPTE-to-MIDI time code, and SMPTE-to-Performer's direct time-lock synchronization code.

Fig. 9-14. The PPS-1 Version Three MIDI sync box.
(Courtesy of J.L. Cooper Electronics)

In FSK mode, the PPS-1 is able to synchronize devices (such as a multitrack tape machine) with MIDI sequencers and drum machines that can recognize SPP messages. This enables the MIDI device to "chase-and-lock" to tape. In the SMPTE mode, this device is able to read all formats of longitudinal SMPTE and convert it into either MTC or DTL sync codes. Additionally, this device can be used as an inexpensive time-code generator for striping time code onto tape.

As with the PPS-1, the PPS-100 from J.L. Cooper Electronics (Fig. 9-15) reads and generates all formats of SMPTE time code, as well as converts SMPTE time code to MIDI sync with SPP, MIDI time code and direct time lock.

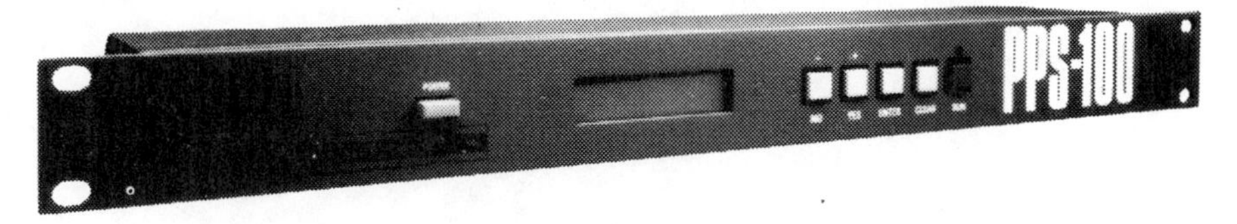

Fig. 9-15. The PPS-100 SMPTE/MIDI synchronizer.
(*Courtesy of J.L. Cooper Electronics*)

Advanced features of this device enable the user to program tempo changes and time signatures into its internal memory, permitting accurate SMPTE to MIDI SPP synchronization within a rhythmically complex composition. While converting SMPTE to MIDI sync, the PPS-100 can generate 24, 48, or 96 pulses per quarter note, in addition to Roland DIN sync (for synchronization with earlier Roland drum machines).

The PPS-100 may also be used as a SMPTE list-based triggering device. The event generator features two independent relay contact outputs (for controlling external switched events, such as a tape recorder's remote footswitch jack, automated effects bypass, and automated muting) and two independent 5-volt pulse outputs (for use with devices that require a voltage trigger). The device's SMPTE-based triggering capabilities also allow for extensive control over MIDI-based instruments, effects, and mixing devices through the generation of SysEx messages.

An optional program, known as PPS-Que, allows tempo maps and events to be entered into an Atari ST or Macintosh computer in the form of a cue list that can be saved and downloaded to the PPS-100. This program may be run alongside many applications as a desk accessory while another program is operating.

The Timecode Machine from Opcode Systems, Inc. (Fig. 9-16) is another SMPTE-to-MIDI converter designed to handle all practical applications which use SMPTE time code to sync audio or video with MIDI and sequencer tracks. These include the ability to read SMPTE time code and output MIDI time code or direct time lock. With the help of a computer, it is also capable of striping SMPTE code onto tape. The Timecode Machine is able to read and write all SMPTE formats, including 24, 25, 30, and drop frame. A Macintosh software desk accessory is also included for easy access to all control commands.

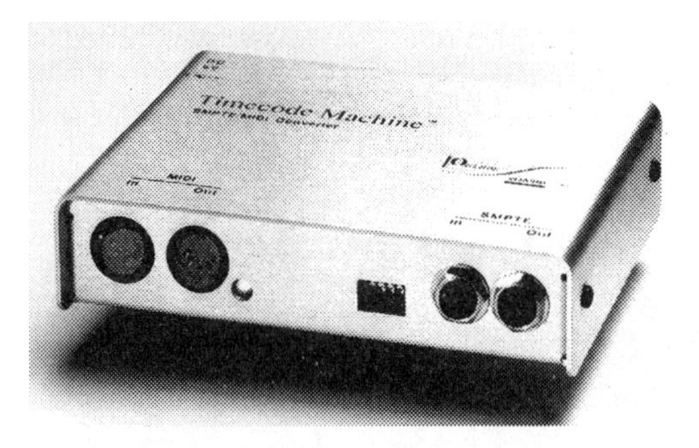

Fig. 9-16. The Timecode Machine SMPTE to MIDI time-code converter. *(Courtesy of Opcode Systems, Inc.)*

Additional features include the ability to jam sync incoming time code, meaning the device is able to generate steady MIDI time code, even if the SMPTE track being read contains dropouts or invalid data. It is also capable of regenerating fresh SMPTE code (in a jam-sync fashion) from a prerecorded time-code source, when dubbing from one audio recorder to another or to a video recorder. The Timecode Machine allows access to SMPTE user bits for user-defined tape identification.

MIDI data entering the device's MIDI in port is allowed to pass through to the MIDI out port, while being merged with MIDI time code. This is useful for devices with only a single MIDI in port since it permits MIDI instruments to be played as time code is being generated.

The Aphex Studio Clock Model 800 from Aphex Systems Ltd. (Fig. 9-17) is another SMPTE-to-MIDI converter capable of reading and writing all SMPTE time-code formats, in addition to outputting MIDI clock, SPP, or MIDI time code. This device makes use of Aphex's Human Clock Algorithm and Rise Time Detection Circuitry, allowing the user to create synchronized tempo maps from such sources as live or recorded drums, bass, keyboards, MIDI clock, MIDI note on, quarter note conductor click, or all of these allowing various players to write the map during different parts of the performance. The Studio Clock also includes such features as jam sync, load/save tempo maps to any MIDI sequencer via Aphex Cache Patch System, and a Macintosh software interface.

Fig. 9-17. The Aphex Studio Clock Model 800 SMPTE-to-MIDI converter. *(Courtesy of Aphex Systems Ltd.)*

Chapter 10

Digital Audio Technology within MIDI Production

Digital audio technology often plays a strong role within MIDI production. This is because MIDI is a digital medium that can easily be interfaced with devices that can output or control digital audio. Devices such as *DAT* (*digital audio tape*) recorders, samplers, sample editors, and hard-disk recorders are employed to record, reproduce, and transfer sound within such a production environment.

Although MIDI is commonly synced to multitrack tape, and final productions commonly make their way to this tape-based medium, the means by which MIDI production systems utilize digital audio is often quite different than those found within the traditional recording studio environment. Whereas the tape-based recording often encodes digital audio as a continuous data stream onto tape, the trigger and control capabilities of MIDI production most often make use of ROM, RAM, and hard-disk sampling technology.

As mentioned in Chapter 4, recordable sample technology allows sounds to be digitized and placed into a storage medium (known as RAM). Once recorded into RAM, these sounds can be mapped across a keyboard or other controller types and finally triggered via MIDI.

In recent years, the way the electronic musician has been able to store, manipulate, and transmit digital audio has changed dramatically. As with most other media, this has been brought about by the integration of the personal computer into the MIDI system for the purpose of digital signal processing and management. In addition to the performance of sequencing, editing, librarian, and notation functions, the computer and related peripherals is also capable of managing digital audio within a MIDI system. This audio can consist of both samples (which may be imported and exported to various sampling devices) and larger sample files on hard disk (providing the artist with digital random access tracks).

The Sample Editor

Over the past few years a number of MIDI sample-dump formats have been developed which enable sample files to be transmitted and received in the digital domain between various types of samplers. To take full advantage of this process for transmitting digital audio within an electronic music system, *sample-editing software* (Fig. 10-1) was developed which allows the personal computer to perform a variety of important sample-related tasks, such as:

- Loading sample files into a computer, where they may be stored within a larger central hard disk, arranged into a library that best suits the users needs, and transmitted to any sampling device within the system (often to samplers with different sample rates and bit resolutions).
- Editing sample files and using standard computer cut-and-paste edit tools to properly arrange the sample before copying to disk or transmitting back to a sampler. Because loops are often supported, segments of a sample are repeated which saves valuable sample memory.
- *Digital signal processing* (*DSP*), allows sample files to be digitally altered or mixed with another samplefile (i.e., gain changing, mixing, equalization, inversion, reversal, muting, fading, crossfading, and time compression).

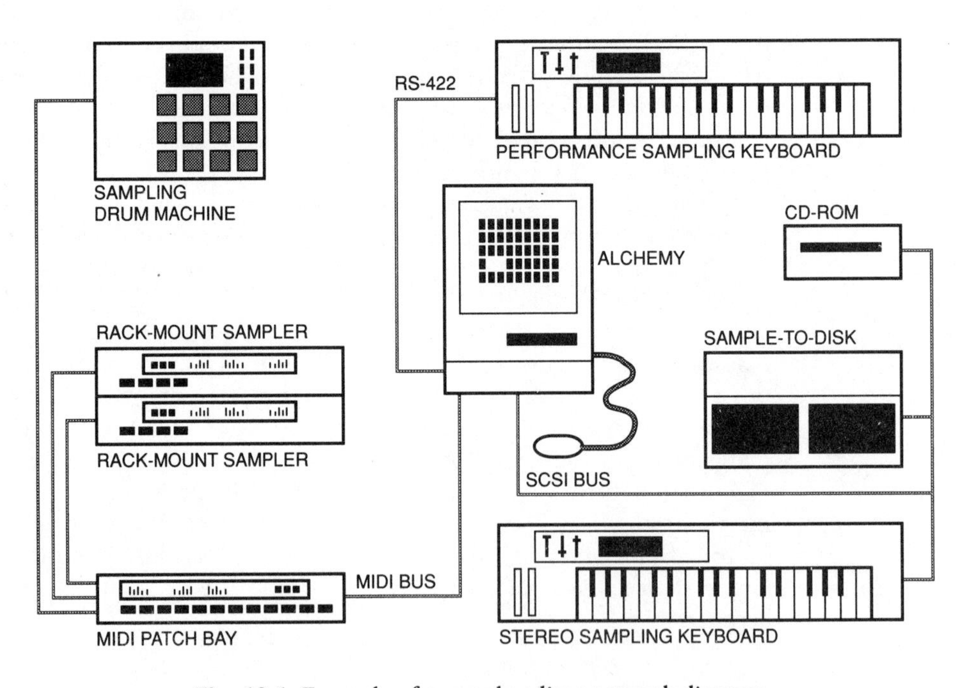

Fig. 10-1. Example of a sample-editor network diagram. (*Courtesy of Passport Designs*)

From these tasks it can be seen that it is basically the role of the sample editor to integrate the majority of sample editing and transmission tasks within an electronic music setup.

The Alchemy Version 2.0 from Passport Designs is a sample-waveform editor for the Macintosh which supports a wide range of samplers. Alchemy is a 16-bit stereo sample-editing network (Fig. 10-2) that allows samples to be imported into a computer via the MIDI sample-dump format and Apple SND resource file format (for integration with Hypercard, Videoworks, and Mac Recorder).

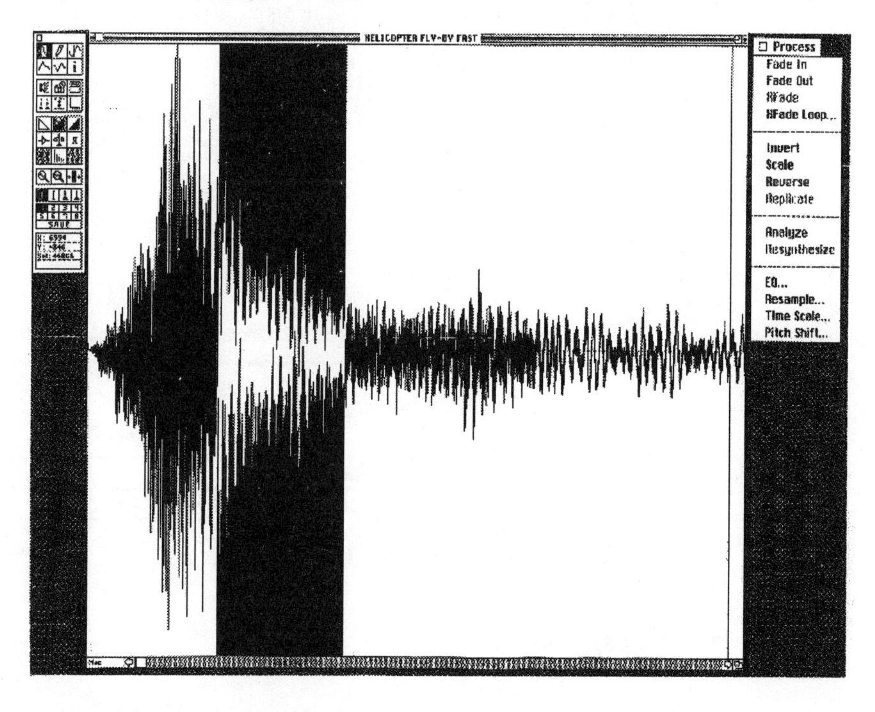

Fig. 10-2. Alchemy main visual-editing software screen. (*Courtesy of Passport Designs*)

Once sample data is loaded into the software, Alchemy functions as a 16-bit sample editor, processor, and network support to many industry-standard samplers. *Fast-fourier transform* (*FFT*) harmonic analysis is included for assistance with harmonic editing and waveform resynthesis. This inclusion allows the user to create new waveforms from original existing samples. Alchemy is also a universal sample translator which allows sample files to be transferred between samplers or distributed systemwide from a centralized sample library.

Another such sample editor for use with an IBM PC/XT/AT or compatible is the SampleVision visual-editing software from Turtle Beach (Fig. 10-3). SampleVision provides a mouse-based, graphic user interface with icons, cut-and-paste sound editing, and pull down menus which are similar to the Apple Macintosh computer.

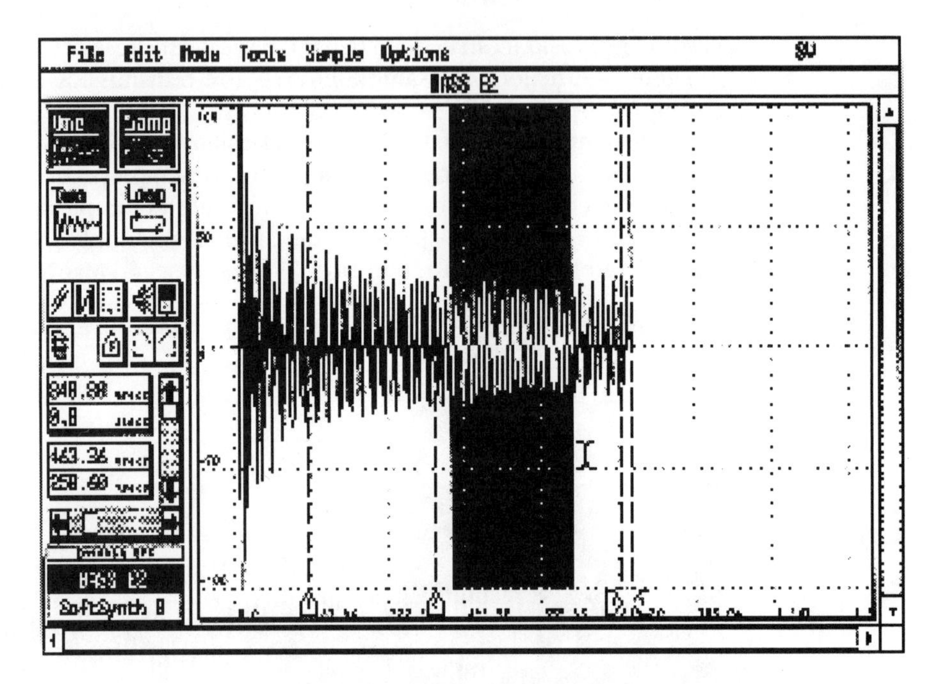

(A) Sample-editing screen.

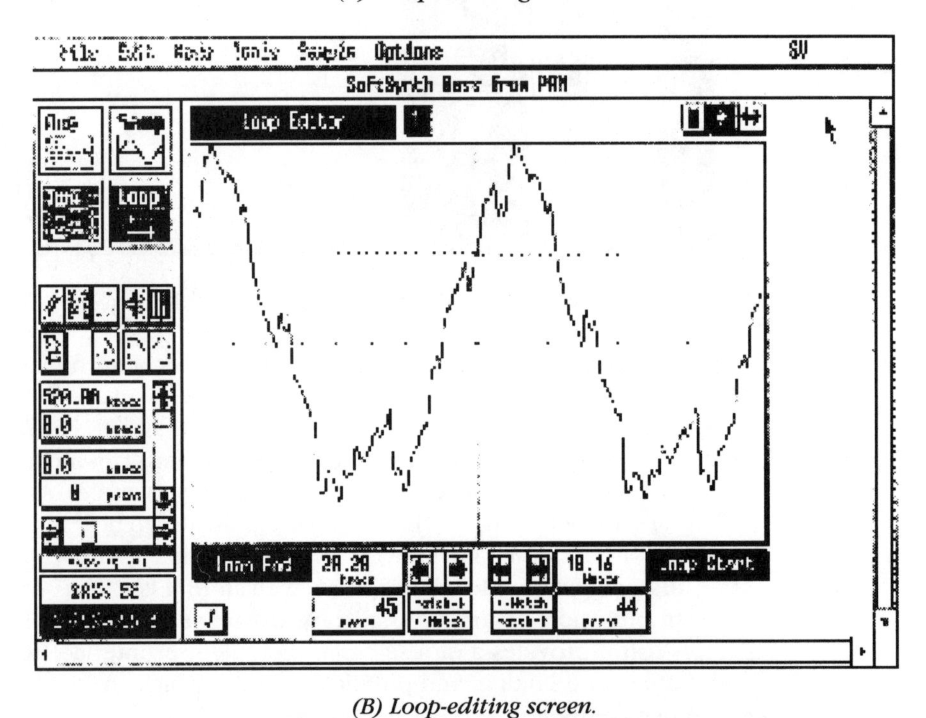

(B) Loop-editing screen.

Fig. 10-3. The SampleVision visual-editing software screens. *(Courtesy of Turtle Beach Softworks)*

Sounds can be enhanced using SampleVision's toolbox of DSP algorithms. These algorithms provide common signal processing functions while keeping the data within the digital domain. These algorithms include: normalize, mix, equalize, inverse, reverse, interpolate, mute and fade, DC level adjustment, and crossfade. All DSP functions are calculated to a 16-bit resolution for increased accuracy.

Many sound-analysis tools are provided within SampleVision. A zoom function allows detailed study of the waveforms, while an animate function displays the sound as it would appear on an oscilloscope. These functions are able to show details which are not discernable using other viewing techniques. The loop editor displays the loop splice and has an autoloop feature, along with a crossfade loop function.

SampleVision has a very comprehensive frequency analysis section which makes use of FFTs for the production of spectral plots of sampled sound (Fig. 10-4).

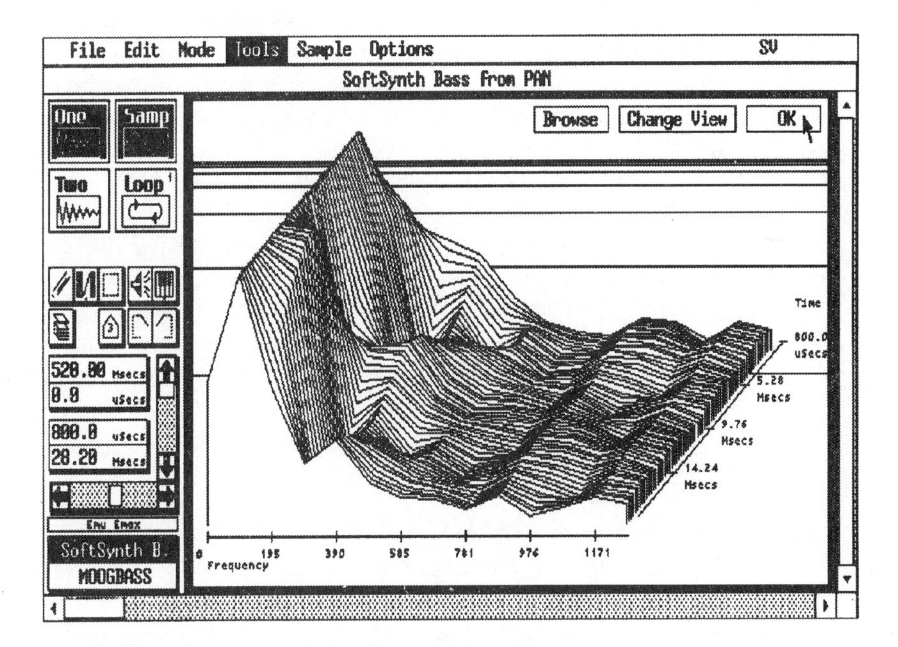

Fig. 10-4. 3-D Fourier frequency analysis screen. (*Courtesy of Turtle Beach Softworks*)

Since optical disks of over 1 gigabyte now exist and high-capacity conventional hard disks are also available, SampleVision allows the editing of large pieces of audio data, with a 32-bit wide data word limit of 4,294 Mb. The screen display can be calibrated by sample count, elapsed time, file percentage, or SMPTE time (supporting all four SMPTE formats).

SampleVision uses a driver architecture to interface with a variety of digital sampling keyboards, such as the Ensoniq EPS, the Yamaha TX16W, the E-MU EMAX, the Akai S900, and any device supporting the MMA sample-dump standard. Drivers are being developed for additional samplers. File level interface is also provided for Sound Designer file formats and for binary 8, 12, and 16-bit sampled data files. Users with multiple samplers can use SampleVision to create a MIDI sampler network with the program acting as the hub, moving samples between them at will.

Hard-Disk Recording

Once the developers had begun to design updated sample-editor software versions, it was found that through additional hardware processing, digital audio editors were capable of recording digitized audio directly to a computer's hard disk. Thus, the concept of the ***hard-disk recorder*** was born. These devices, occasionally also known as ***digital audio workstations***, serve as computer-based hardware and software packages which are specifically intended for the recording, manipulation, and reproduction of digital audio data which resides upon the hard disk and/or within the computer's own RAM. Commonly, such devices are designed around and controlled by a standard personal computer (such as, a Macintosh, IBM, or Atari) and involve the use of an additional proprietary digital co-processor. Such a co-processor is required since the speed and number crunching capabilities of the PC's processor alone is often not enough to perform the complex DSP calculations that are encountered within digital audio production. Other more expensive multichannel systems often make use of propriety computer and processing systems specifically designed to perform digital audio and signal processing tasks.

The advantage of such a workstation within a MIDI production environment is multifold. Such capabilities include:

- *Longer sample files*: Hard-disk recording time is often limited only by the size of the hard disk (Commonly 1 minute of stereo audio at 44.1 kHz will occupy 10 Mb of hard-disk memory).
- *Random access editing*: While audio is recorded upon a hard disk, any point within the program may be accessed at any time, regardless of the order in which it was recorded. Additionally, nondestructive hard-disk editing may be used, whereby defined segments of audio may be placed in any order within a cuelist or playlist and reproduced from this list in either consecutive order or at specified SMPTE time-code addresses.
- *DSP*: Digital signal processing may be performed upon a segment or entire samplefile, in either real or nonreal time (i.e., equalization, mixing, gain changes, reversal, or resynthesis).

In addition to the above advantages, this computer-based digital audio device functions to integrate many of the tasks related to both digital audio and MIDI production. Like the sample editor, hard-disk workstations are often capable of importing, processing, and exporting sample files. Thus, such a device offers a new degree of power to the artist that relies upon sampling technology. It is also capable of acting as a samplefile librarian, editor, and signal processor.

An example of a popular 2-channel digital audio workstation is the Sound Tools desktop audio-production system from Digidesign. This system (Fig. 10-5), which is designed for use with the Macintosh II, SE, and Atari personal computers, consists of three elements: the Sound Accelerator digital signal processing card, AD IN analog to digital converter, and Sound Designer II audio-editing software. Together, these products form a stereo hard-disk recording system that provides the user with a wide range of record, edit, and DSP capabilities.

Fig. 10-5. Sound Tools desktop audio-production system. (*Courtesy of Digidesign, Inc.*)

The Sound Accelerator digital signal processing card is used to output 16-bit stereo audio from a Mac II series or SE computer. Through the use of a Motorola 56001 processing chip, the Sound Accelerator is capable of performing such complex DSP functions as time compression/expansion and parametric and graphic EQ. In addition to Sound Tools, this single card may be used with a multitude of sound-related programs for the reproduction of 16-bit stereo audio, including Turbosynth, Sound Designer II, and Softsynth.

The AD IN analog-to-digital converter operates in conjunction with the Sound Accelerator to record any line-level stereo or mono-audio signal, directly to a Mac hard disk. The AD IN is optimized for a 44.1 kHz sampling rate. However, when recording in mono, a 2-times oversampled rate of 88.2 kHz is also available. Front panel controls include: power, stereo level adjustments, stereo/mono-mode selector, and LED signal/overload indicators.

Direct digital audio communication is available through the use of Digidesign's DAT I/O (pronounced *daddy-oh*) digital interface. Digital audio data may be received and transmitted from the Sound Tools system with no degradation in audio quality via the AES/EBU professional digital 2-channel communications bus and the S/P-DIF consumer DAT and CD interface formats.

The Sound Designer II audio-editing software (Fig. 10-6) features five primary areas of operation: recording and editing, digital signal processing, stereo sample editing, time-code synchronization, and sample networking within a MIDI system. The following represents only a brief introduction to the system's hardware and software capabilities.

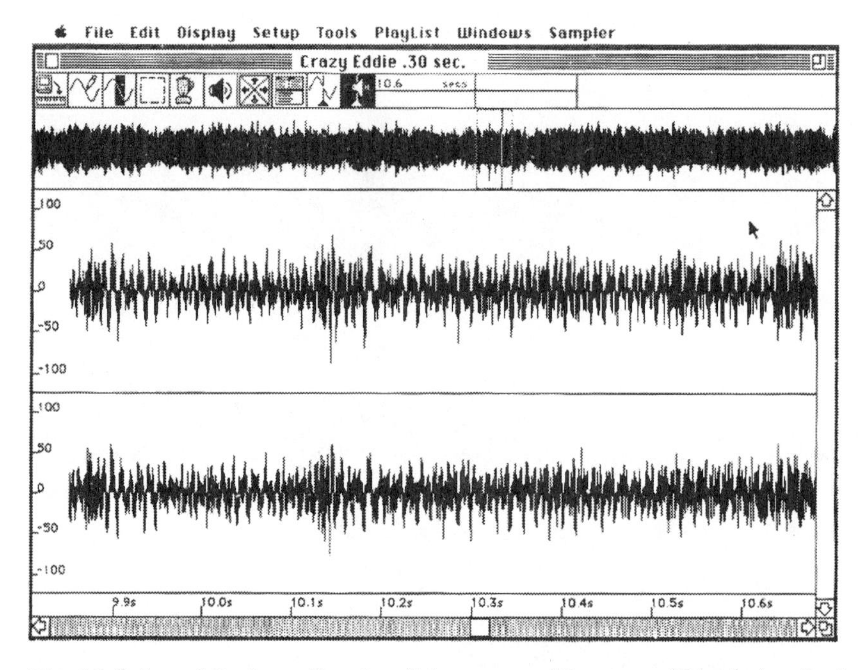

Fig. 10-6. Sound Designer II main editing screen. (*Courtesy of Digidesign, Inc.*)

All record functions may be accessed through the tape deck module, which has been designed to mimic the functions of a reel-based recorder. Audio inputs can be monitored from the system's outputs in a throughput fashion, while the signal is graphically monitored using onscreen, multisegment VU meters. Sampling may be done at any user specified rate. However, the system is optimized for recording at 44.1 kHz.

Since Sound Tools is primarily a hard-disk workstation, most edit functions are carried out through the use of nondestructive playlist editing. This form of editing does not affect a recorded soundfile in any way, but simply defines a region as a set of markers and parameters that instruct the system as to which part of the soundfile to reproduce and how to reproduce it. Using this approach to editing, once the desired number of regions has been selected and identified from an overall soundfile, these soundbites may be positioned in any order within a playlist (Fig. 10-7) for sequential reproduction or trigger to time code.

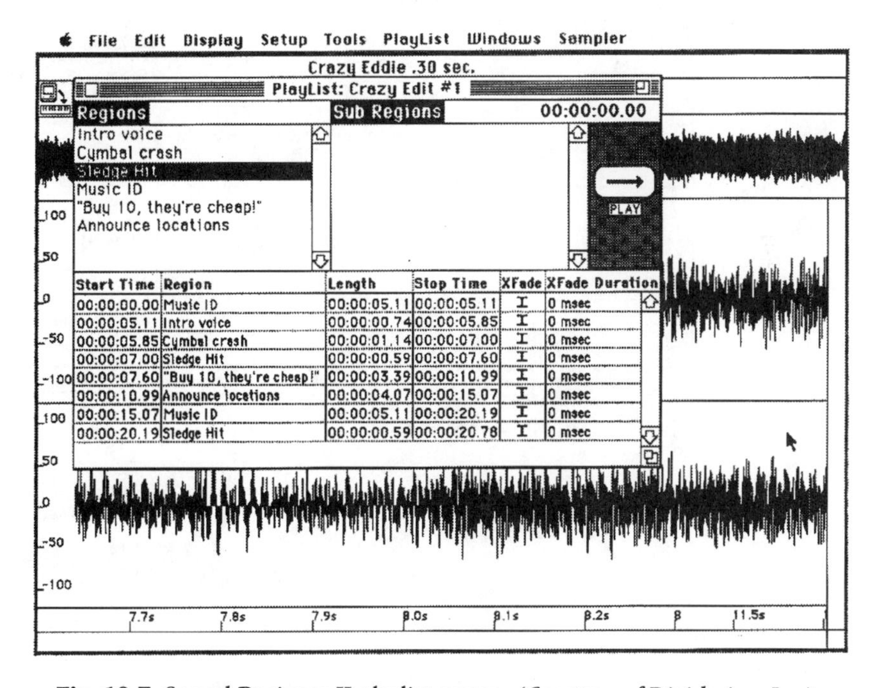

Fig. 10-7. Sound Designer II playlist screen. *(Courtesy of Digidesign, Inc.)*

Sound Designer offers such signal processing capabilities as mix, crossfade, EQ, time compression, and expansion. EQ may be placed into the signal path or added to a region in real time through the use of a 7-band graphic or parametric equalizer offering control over center frequency, bandwidth, and gain for each of 7 bands per channel.

Synchronization between Sound Tools and any device employing SMPTE time code is accomplished through the use of MIDI time code. This form of synchronization allows a playlist region to be triggered at a precise time-code address. This allows regions to be synchronously triggered from another time encoded device, such as a video tape recorder or SMPTE-based sequencer package.

In addition, Opcode Systems and Digidesign have collaborated to create a powerful new software package that integrates the MIDI sequencing power of Opcode's Vision with the digital audio playback and editing capabilities of Sound Tools. This program, known as Studio Vision (Fig. 10-8), allows a Macintosh SE or MAC II to record and manipulate two tracks of CD-quality audio with full simultaneous MIDI sequencing capabilities. Additionally, Studio Vision makes use of Vision's MIDI controller faders and other mix controls to bring automated mixing to the digital tracks.

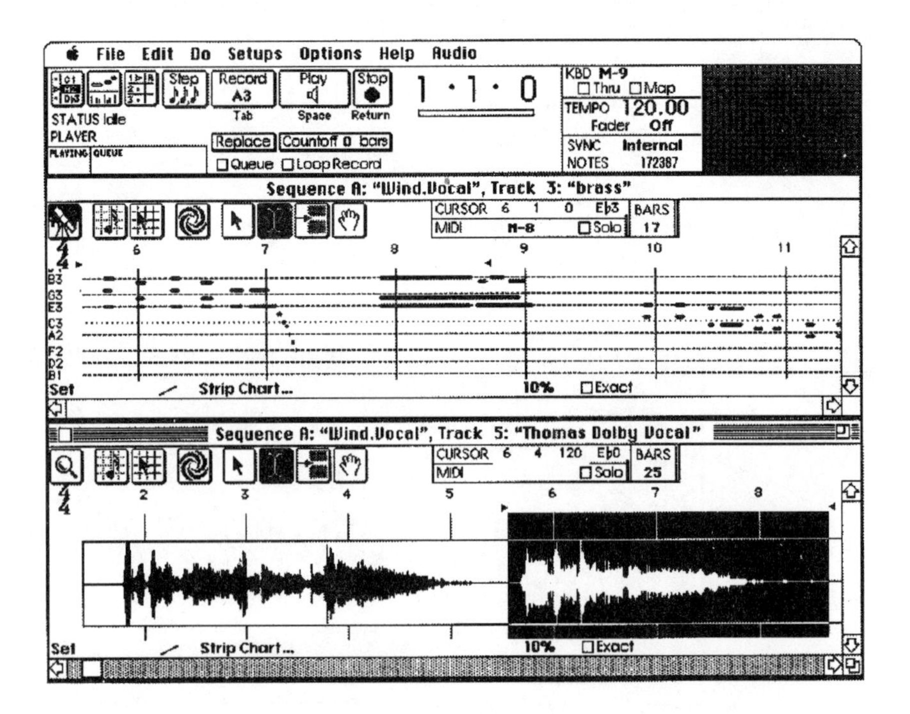

Fig. 10-8. Main screen of Studio Vision, showing MIDI and digital audio tracks. *(Courtesy of Opcode Systems, Inc.)*

Synthesis and Sample Resynthesis

In addition to computer-based sample edit and signal processing functions, software packages make use of various MIDI sample-dump formats to import a sample. Once imported, such a program is then capable of performing extensive sound synthesis and resynthesis (the use of computer algorithm to change an existing sound) upon that sample. Afterwards, it can be used to transfer the newly regenerated sound back to a sampling device or hard-disk recorder.

Turbosynth from Digidesign (Fig. 10-9) is one such modular-synthesis and sample processing program which combines the elements of digital synthesis, sampling, and signal processing for the creation of sounds with a Macintosh Plus, SE or II series, and any sampling system.

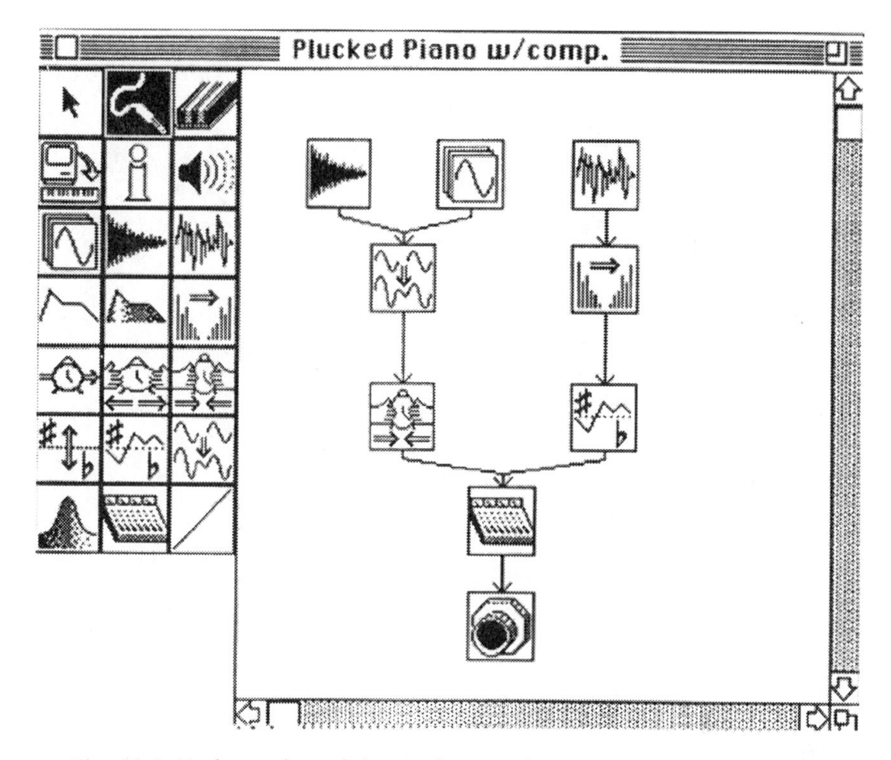

Fig. 10-9. Turbosynth modular-synthesis and sample processing program. (*Courtesy of Digidesign, Inc.*)

Modular synthesis works by combining various modules into a sound patch to generate or modify a sound. Turbosynth's modules include oscillators to generate basic waveforms, filters to modify the harmonic content of the oscillator's signal, and amplifiers to vary the volume of the sound. In addition to these traditional analog style modules, Turbosynth includes a number of digital processing modules which do not exist in analog-modular synths for the production of a wider range of sounds. Once created, a sound patch can be saved to disk for later recall or transfer to any of the currently available samplers for full polyphonic reproduction.

Any soundfile created in Digidesign's Sound Designer format may be used as a sound source within a Turbosynth patch. This allows a sampled soundfile to be digitally modified to create new, wild, or exotic sounds. As with any sound patch, a modified sample may be saved to disk and/or transferred back to a sampler.

Softsynth from Digidesign (Fig. 10-10) for the Macintosh and Atari computers makes use of the MIDI sample-dump standard to turn virtually any sampler into an additive or FM digital synthesizer.

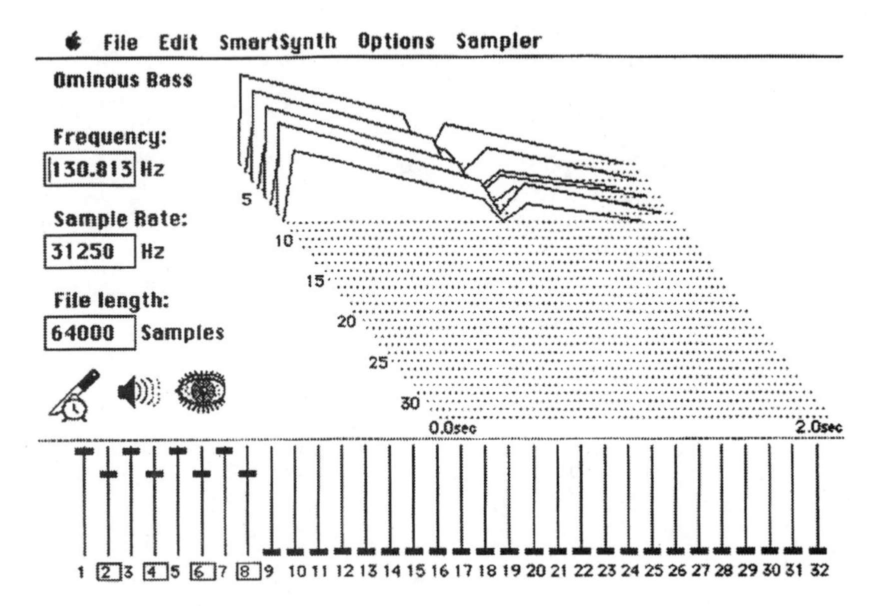

Fig. 10-10. Softsynth additive or FM digital synthesis program (showing time-slice screen). (*Courtesy of Digidesign, Inc.*)

Each of Softsynth's 32 variable frequency oscillators has a 40-stage amplitude envelope, a 15-stage pitch envelope, and a choice of five different waveforms. Once the desired 16-bit sound has been designed using graphic programming screens, it may be transferred to a sampler for playback.

Distribution of Sampled Audio

Within a sample-based MIDI setup, it is important that the distribution of sampled audio be as fast and as painless as possible. Most commonly, samples are managed as either samples (encoded digital audio data) or sample files (computer files containing encoded digital audio data). In order for most samplers and related software management programs to communicate systemwide, guideline standards have been adopted for the transmission and storage of digital audio.

MIDI Sample-Dump Standard

The *sample-dump standard* (*SDS*) was developed and proposed to the MIDI Manufacturers Association by Chris Meyer and Dave Rossum as a protocol for transmitting sampled digital audio and sustain-loop information from one sampling device to another. This data is transmitted as a series of MIDI SysEx messages. However, SDS is a data protocol for the successful transfer of these messages between sampling hardware.

Although most samplers share common characteristics, many sampling devices may have their own unique set of SysEx messages. This requires the involved samplers to share a common SysEx structure or that a program (such as a sample editor) be used that is capable of communicating with a wide range of sampler types.

One of the greatest drawbacks of SDS is that data transfer can be a rather slow process, since it conforms to the MIDI standard of transmitting serial data at the 31.25-kbaud rate.

SCSI Sample-Dump Formats

Certain computer-based digital audio systems and sampling devices are capable of transmitting and receiving sampled audio via a *SCSI (small computer systems interface)*. SCSI is a bidirectional communications bus commonly used by many personal computer systems to exchange digital data between systems at high speeds. When used in digital audio applications, it provides a direct data link for transferring soundfiles at a rate of 500,000 bits/second (nearly 17 times MIDI speed) or higher. Such an interface, when combined with a computer-based digital audio system, provides the user with a fast and straightforward means of transferring data to and from an editing program and central hard-disk sample library.

Computer Samplefile Formats

Although a number of formats exist for saving samplefile and soundfile data onto computer-based storage media, several popular formats have been adopted by the industry. This allows sample files to be exchanged between compatible sample-editor programs.

Two of the most commonly encountered samplefile formats are Sound Designer and Sound Designer II, which were developed by Digidesign. These formats are both 16-bit linear. However, there are definite distinctions between the two. Sound Designer is an earlier monaural version that is capable of encoding two sample loops (sustain and release) within its structure. The more commonly encountered, recent version, Sound Designer II, is a stereo samplefile and soundfile (hard-disk recording) format capable of encoding up to eight loops within a file.

The *Audio Interchange File Format (Audio IFF)*, conforms to the EA IFF 85 Standard for Interchange Format Files. It allows both monaural and stereo-sampled sounds to be stored with a variety of sample rates and sample widths. Each sample point is processed as a linear 2's complement value (a straightforward method of storing positive and negative sample values) with a possible bit range from 1- to 32-bits wide. Sound data can be looped within the Audio IFF format, allowing a portion of the sound to be repeated in either a forward loop or forward/backward loop.

Memory Management

One of the central issues surrounding RAM-based digital audio is memory (or the constant need for more of it). Since approximately 10 Mb of data is required to store one minute of stereo digital audio, it is easy to see how memory capacity can quickly become an issue.

Random Access Memory

Initially, most sample-based digital devices make extensive use of internal RAM to process raw digital data. Thus, the length and number of sounds a sampler may record or output is in part determined by its internal RAM. This also holds true for many computer-based programs, since all or part of their signal processing is often performed in RAM. For this reason, it is highly recommended that a system have a sufficient amount of RAM or may be easily upgraded to fulfill future requirements.

Magnetic Media

An inexpensive and common method for storing computer-based digital data is using magnetic media, such as a floppy and hard disk.

The *floppy disk* (also known as a *diskette*) is a low-density medium for the recording, storage, and reproduction of digital data. These diskettes are most notably available as 5 1/4-inch–360-Kb/1.2-Mb *floppies* and 3 1/2-inch[en]720/800-Kb and 1.4-Mb *stiffies*. Common uses for these disks include storing program data, text, and graphics. However, due to their limited storage capacity, only a few seconds of audio data can be stored onto a single disk.

The *hard disk* is a high-density, high-capacity medium for the storage and retrieval of data. These devices, which can be housed either within a computer-based device or externally as a stand-alone unit, are made up of rotating, magnetically recordable disks, which contain a read/write head that floats over the magnetic surface on a thin cushion of air. Due to their fast search times and higher data densities (ranging from 20 Mb to over 1 gigabyte) these devices are currently among the fastest and most cost-effective memory devices available.

Hard disks are available in three basic types: internal, external (luggables and portables), and removable. Removable hard disks are comprised of an external drive system and hard-disk cassettes. This is a fairly robust medium with general hard-disk capacities being in the 40–45 Mb range.

Optical Media

Although the storage of digital data onto an *optical disc* (the compact disc) is one of the greatest success stories in recent decades, other forms of optical storage are becoming increasingly familiar to multimedia producers. These include the CD-ROM, WORM, and erasable optical media.

The process of encoding digital data onto an optical disc makes use of low-level laser technology to encode digital data directly onto a rotating disc. This has the effect of directly changing the disc's surface either chemically or optomagnetically to produce a stream of digitally encoded pits or nonreflective surface areas.

Optical data storage devices offer distinct advantages over hard-disk memory since they are capable of offering much greater data densities (with storage capacities often being measured in the gigabyte range) at a fraction of a hard disk's cost per megabyte. Additionally, optical media has the definite advantage of being removable. This circumvents the dreaded *disk-full syndrome*. Since the medium is robust, it often reduces the need for data transfer to a backup medium.

The *Compact Disc* (*CD*) offers both the professional and consumer markets a removable random-access 16-bit digital audio medium. This mass-produced read-only format offers up to 74 minutes of playing and has an overall data capacity of about 15.6 billion channel bits of information, of which about 625 Mb is made up of audio data.

In addition to digital audio, the CD format can be used to store computer-related data, as is the case with the *CD-ROM* (*compact disc/read-only memory*) *format*. Unlike the audio compact disc, CD-ROM is not tied to any specific application. In addition, it can be used to access up to 650 Mb of computer-related data in the form of text, graphics, video still pictures, and even audio samples.

Optical drives are also available, which allow data to be written (encoded) onto disc. This format, known as a *WORM* (*write-once read-many*) system, allows data to be written to disc only once. Thereafter it is read an infinite number of times. CD-recordable drives are also becoming available. They allow data to be written to a CD-type disc that conforms with the CD standard.

The most recent addition to the optical data-storage media is the *erasable optical disk*. This medium offers the dual advantages of high-data density with the flexibility of erasing and re-encoding data. Currently, such systems generally transfer and verify data to disk at a rate too slow for real-time applications, such as hard-disk recording. However, as soon as speed, price, and standardization obstacles are transcended, this medium will most likely have a great impact upon digital audio in general.

Digital Audio Tape

Although digital audio tape (DAT) is generally not considered to be a direct method for encoding computer-related data (although certain manufacturers are currently tackling the storage of samplefile-related data to DAT), it is a cost-effective means for obtaining and archiving your own samples with digital quality. It is an extremely useful medium within MIDI production, since soundfiles can be transferred to DAT (eliminating the need for an extremely large hard disk or extensive backup facilities). In addition, partial or final mixes may be recorded or transferred to DAT for mastering purposes with digital CD quality.

Chapter 11

MIDI-Based Mixing and Automation

Throughout the last decade MIDI has changed the face of popular music production, and brought about major changes in design and technique within the modern recording studio. One of the last bastions of the professional studio, however, is often considered to be the professional mixing console and the expertise that sits in front of it. This is one of the major points which marks a perceived dividing line between the electronic-music and professional-recording market.

In the past, almost all commercial music was mixed by a professional recording engineer under the supervision of a producer and/or artist. With the emergence of electronic music, high-quality production has become more cost-effective and easily reproduced. MIDI workspaces can now be owned by individuals or small businesses which are quite capable of producing professional sounds. Many such systems are often operated by artists who are experienced in the common sense rules of creative and commercial mixing with final results that speak for themselves.

Most professional mixers had to earn their "ears" by logging in many hours behind the console. Although there is no substitute for this expertise, the mixing abilities and ears of electronic musicians are also improving with time as they become more knowledgeable about proper mixing environments and techniques by logging in their own hours and mixing their own compositions.

Nonautomated Mixing within Electronic Music

The field of electronic music is also affecting the requirements of mixing hardware. This is due to the need for control over the large number of physical inputs, outputs, and effects which are commonly encountered within many MIDI production facilities.

Although traditional console design has not changed significantly in recent years, electronic music production has placed new demands upon these devices. For example, a synth with 4 outputs, a drum machine with 6 outputs, and sampler with 8 outputs might dominate an average mixer, leaving room for little else. In such a case, a system might easily outgrow a console's capabilities, leaving you with the unpleasant choice of either upgrading or dealing with your present mixer as best you can. For this reason, it is always wise to anticipate your future mixing needs when buying a console or mixer.

One popular method for handling the increased number of inputs is to make use of an outboard line mixer (Figure 11-1 and 11-2). These rack-mountable line- or sub-mixers are commonly able to handle up to 16 or 24 line-level inputs, which are then mixed down to two channels. These two channels can then be routed to either two inputs, auxiliary, or monitor returns on your main mixing system. On some designs, they can be routed to special multi-inputs which allow multiple mixers to be directly tied together without using additional line-level input strips. Often, newer rack submixers include facilities for effects sends and returns.

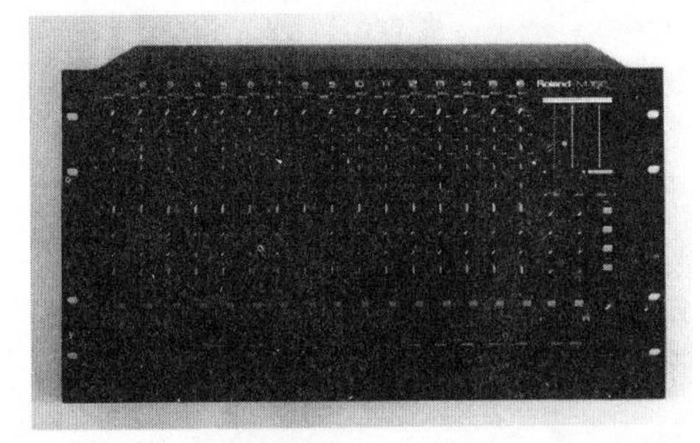

Fig. 11-1. The Roland M 16E line mixer. (*Courtesy of Roland Corporation US*)

Fig. 11-2. The Rane SM 82 stereo 8-channel line mixer. (*Courtesy of Rane Corporation*)

Mixing Via MIDI

Although standard mixing practices still predominate within most multitrack recording studios and many MIDI production studios, MIDI itself also provides the artist with the capability to perform cost-effective mixing and signal control.

This can be accomplished by altering MIDI channel messages so the data pertaining to an instrument or device's dynamic levels, panning, etc. can be directly imbedded within the sequence.

Alternatively, newer generations of MIDI hardware and software systems are appearing on the market. These systems are able to provide extensive mixing control over musical instruments, effects devices, and/or dedicated MIDI-based audio mixers through the use of dedicated MIDI controller messages (Fig. 11-3).

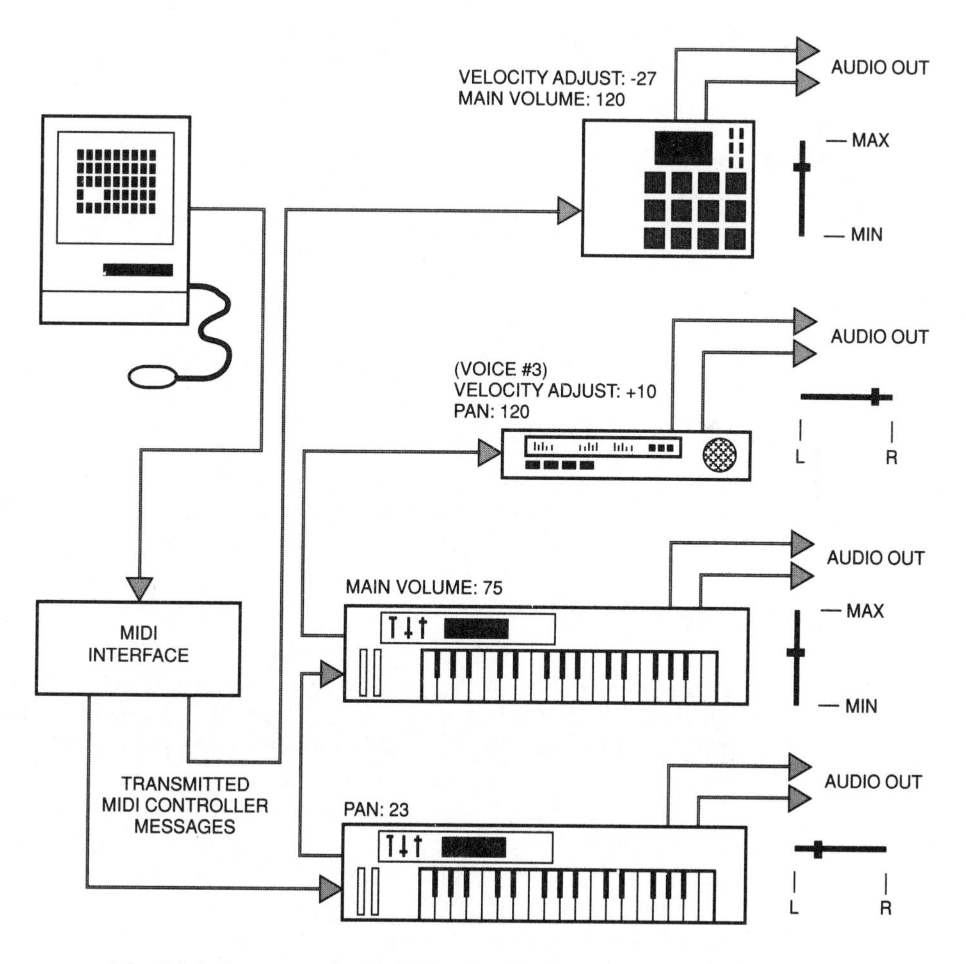

Fig. 11-3. System-wide MIDI-based mixing via MIDI channel messages.

When combined with the power that MIDI provides over music voicing, timbre, effects program changes, etc., MIDI-based mixing and control automation gives the electronic musician a degree of automated mixing control which is, in certain respects, unparalleled in audio history.

Dynamic Mixing Via MIDI Control-Change Messages

The vast majority of electronic instruments allow for dynamic MIDI control over such parameters as velocity (individual note volume), main volume (a device's master output volume), and pan position (of one or more voices). This is accomplished through user control over MIDI voice messages (Fig. 11-4), such as velocity and continuous controller, which can be transmitted to one or more devices as a stream of messages with values ranging from 0 (minimum) to 127 (maximum). As these messages can be transmitted over a series of individual MIDI channels to provide control over individual voices or a series of grouped voices, a form of *MIDI mixing* can be performed upon a composition.

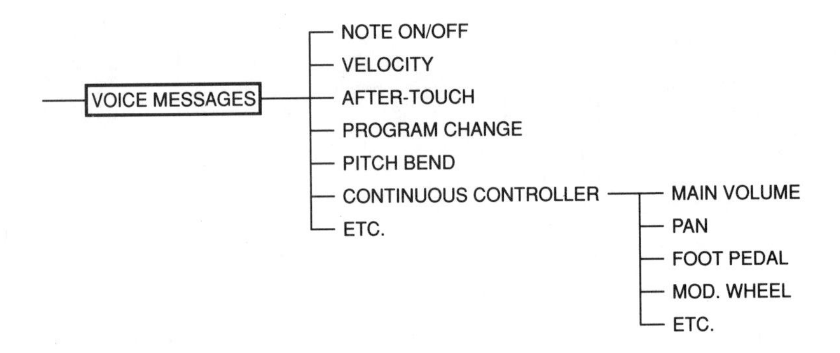

Fig. 11-4. Diagram of many of the voice messages found within the MIDI data stream.

Velocity Messages

Although individual notes can be dynamically played to provide changes in level and expression, many sequencers also give the user control over level within an entire sequenced track or range of notes within a sequence. For example, let's assume that within a computer-based sequence (Fig. 11-5) we have a series of tracks, and that each track has been assigned to a MIDI channel and an associated voice. If we wish to change the gain of an entire track, portion of a track, or range of tracks, it is generally a simple matter to adjust the relative or absolute (equal velocity) levels by selecting the region to be affected and instructing the program as to the desired velocity and method of gain change.

Thereafter, at the selected measure, the proper gain changes will be automatically set. The following examples are a few of the gain-change parameters which are found within Voyetra's IBM/compatible-based Sequencer Plus program.

- *Set velocity*: This gives all notes within the selected range the same MIDI velocity setting.
- *Adjust velocity*: This proportionately increases or decreases the velocity settings of all notes in the range of measures.
- *Crescendo*: This sets the velocity data for all notes within a range of measures, making a gradual increase/decrease in volume from note to note. Although this is not a true crescendo, the volume of the track increases (or decreases) over a period of time.

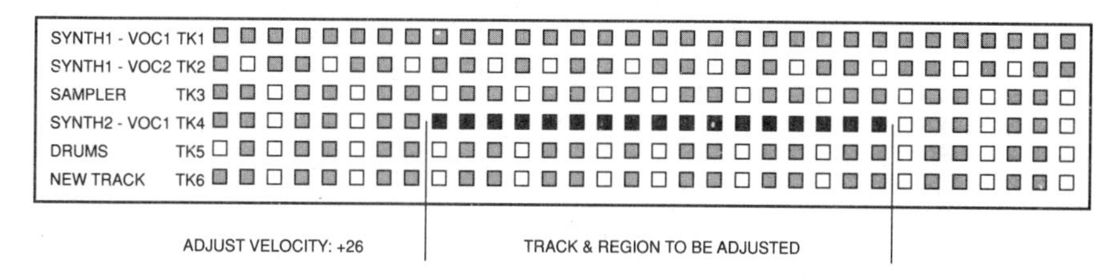

Fig. 11-5. Basic sequencer screen showing region to be gain changed.

Although control over velocity is often used to scale volume, under certain conditions there may be undesirable effects. These could occur should velocity levels also affect other sound parameters, such as the filter cutoff of a synthesized voice or sample trigger points.

Continuous-Controller Messages

As with the use of velocity to control the levels of a sequenced track, controller messages are also able to exert control over dynamic mixing events (Fig. 11-6). One such dynamic controller event, known as *main volume*, is used to vary the overall volume of an instrument, or alternately the patches within a polytimbral instrument. This message can also be used for level control over a wide range of devices, such as the individual gain adjusts for an electronic instrument, output levels of an effects device, or the channel and master gain of a MIDI-controlled mixer.

As has been covered, the MIDI specification makes provisions for 128 different types of controller messages, of which a few controller numbers have been adopted as de facto standards by many manufacturers (i.e., #1—modulation wheel, #4—foot pedal, #7—main volume, #10—pan position, etc.). A range of controller messages (64–95 and 122) have been designated by the MIDI specification as *switches*. These may be set to either 0 (off) or 127 (on) and nothing in between.

For devices equipped with stereo outputs or multiple outputs which can be grouped together into stereo pairs, MIDI pan-controller messages (Fig. 11-7) can be used to pan individual voices or groups of voices from left (value 0) to right (value 127).

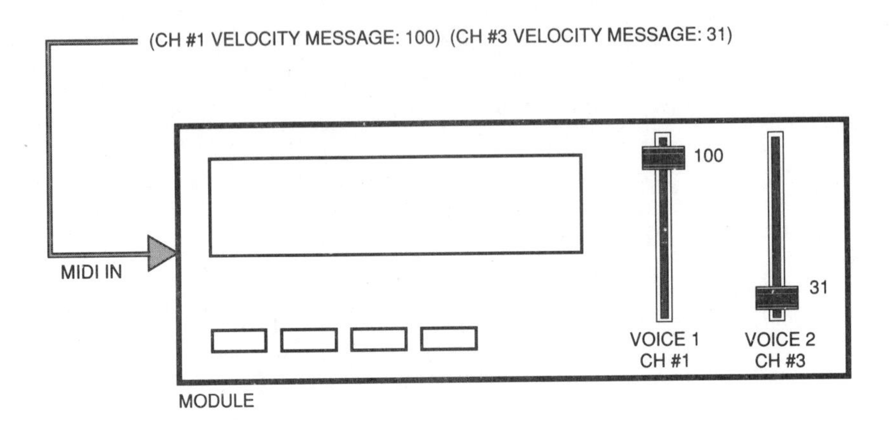

Fig. 11-6. Main-volume controller messages within a MIDI system.

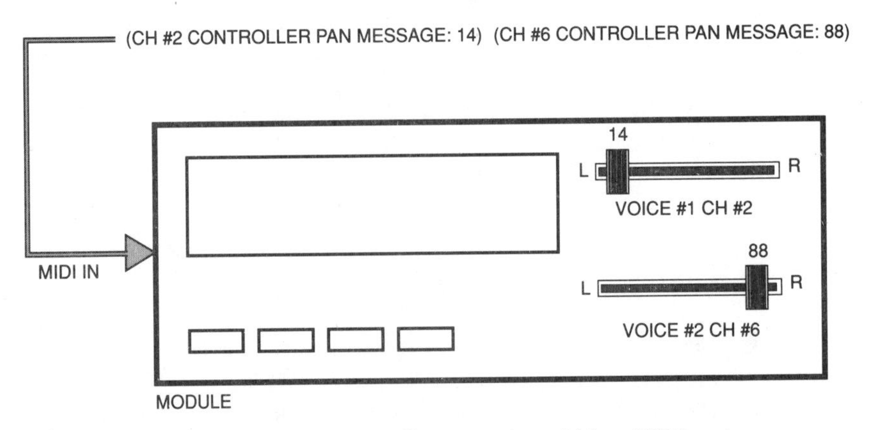

Fig. 11-7. Pan-position controller messages within a MIDI system.

A sequencing device will often have its own method for accessing MIDI controller messages, and certain MIDI devices may make use of nonstandard message numbers for exerting control over dedicated parameters. For these reasons it's always wise to consult the operating manual of a device since it will generally include operating instructions and message data tables.

The Dynamically Mixed Sequence

Unfortunately, since few MIDI setups make use of the same or even similar equipment, the levels of a dynamically mixed sequence may not be totally valid when played back upon another MIDI system. Should you wish to change the gain parameters to match those of a new setup, it is often wise to save the sequence under another filename to retain the gain settings of the alternate MIDI system. Another *fix* might be to manually change the levels at the mixer or console.

Should a completed sequence be prepared for transfer from electronic instruments directly to a multitrack tape about the professional studio, it is highly advisable that the sequence not be mixed in the MIDI domain (e.g., using volume and pan-controller messages). This precaution provides the engineer and all those concerned about the mixdown phase with a greater degree of manual level control at the recording console and multitrack recorder. Such traditional mixing techniques offer an improved signal-to-noise ratio that results from being able to mix instrument outputs which were recorded and are being reproduced at the hottest possible signal level. Additionally, the dubbing of acoustic, electronic, and vocal parts over a sequence will often create needs for changes in the overall balance. Again, should you wish to mix the sequence in the MIDI domain when working at home, before going into the studio you might consider saving the mixed and unmixed version under a different filename.

MIDI Remote Controllers

One of the major drawbacks to mixing velocity and controller messages within the edit screen of certain sequencing programs, is the lack of resources for changing these gain and pan functions in real time. One possible solution to this dilemma is to use a MIDI remote controller (Fig. 11-8).

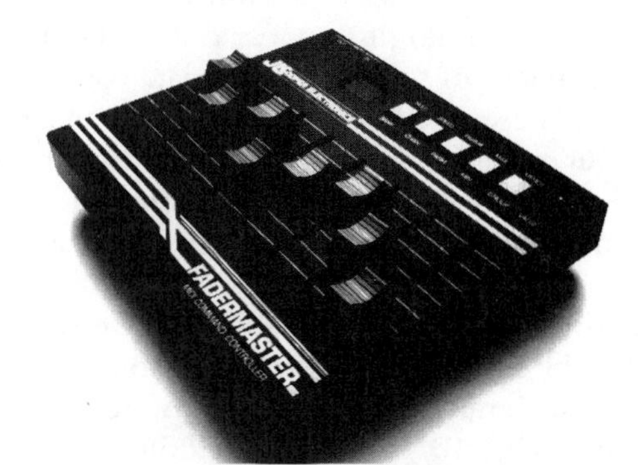

Fig. 11-8. The FaderMaster MIDI remote controller. *(Courtesy of J.L. Cooper Electronics)*

The advantage of using a remote controller to effect dedicated channel messages is its ability to output MIDI messages dynamically in real time by moving a data slider (Fig. 11-9). Such a controller could be programmed to effect continuous gain change, pan messages, etc. over a number of channels or groups of channels. Often such controllers offer up to eight data faders, which can each be dedicated to one or more channels and provide control over one or more channel messages. Such controllers often provide buttons for on/off controller events.

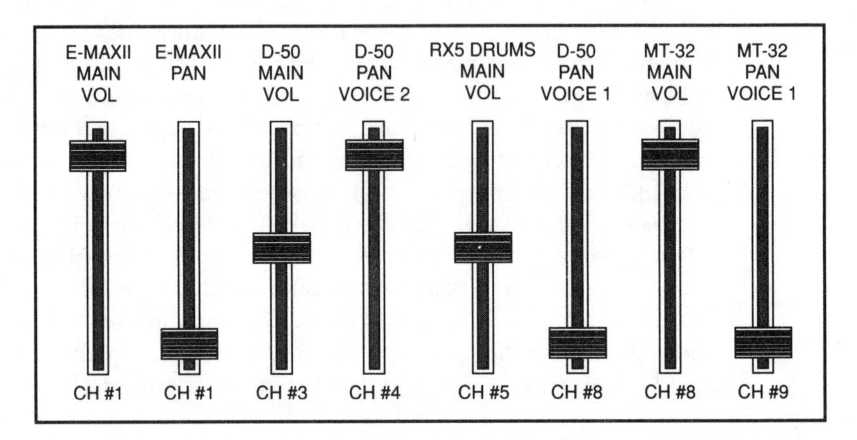

Fig. 11-9. Real-time control over velocity and controller messages is possible through the use of remote controller data faders.

Controller Automation Software

To take better advantage of the full 128 controller messages which are available, software is beginning to appear on the market that allows access to a much greater number of software-based data controllers (such as faders, knobs, and buttons). These controls can be pasted onto a screen and assigned to a specific controller in a number of ways that best fits the application at hand. For example, a program might allow a set of fader controls to be placed within the same screen as the main sequencer page. This would enable the user to vary level, pan, or other related functions in real time while the sequencer is being played. Alternatively, a MIDI-controllable mixer might be remotely automated from a program by laying out a number of control icons onto a screen in a graphic fashion which mimics the entire set of mixer controls.

One such MIDI-based automation system is Q-Sheet A/V, a SMPTE/MIDI automation program from Digidesign. Using appropriate MIDI equipment and a Macintosh computer (minimum RAM of 512K and 800K of disk space), Q-Sheet A/V is able to automate many of the production aspects within a MIDI-based system, including effects units, keyboards, samplers, and MIDI mixing consoles; all locked MTC/SMPTE time code.

Q-Sheet A/V is built around the concept of a user-created cue list which consists of a consecutive series of MIDI events. An event might be the real-time movement of a fader, the adjustment of a delay setting, or simply a note-on/off command for a synthesizer or sampler.

In addition to these capabilities Q-Sheet's integrated graphics and MIDI controller permits the creation of a fully customized automation window (Fig. 11-10). This window allows onscreen faders, knobs, switches, and counters to be easily created and assigned to any MIDI controller parameter. These icons can then be used to directly control related mixing functions by simply moving the mouse's cursor to the icon, clicking the mouse button and moving the control in real time.

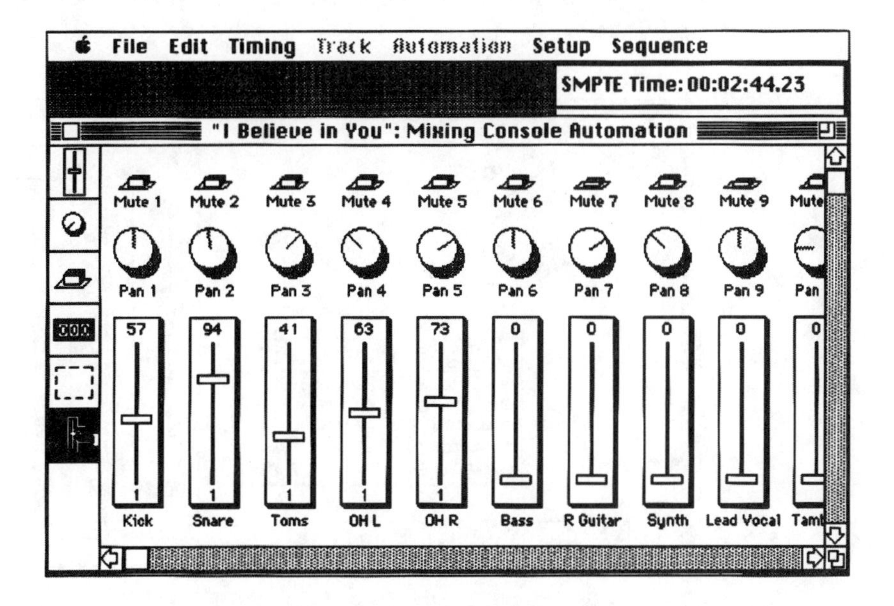

Fig. 11-10. Q-Sheet A/V mixing-console automation window.
(Courtesy of Digidesign, Inc.)

These controllers can easily be linked together, allowing the user to group any number of faders, knobs, or switches. Different controller types may also be linked together. For example, a fader can be linked to a pan pot so that when the level is reduced, the signal will pan from right to left (and vice-versa). Controllers can also be inversely linked. That is, when one fader is lowered, another fader (or grouped faders) will be proportionately raised.

These onscreen controllers can be recorded into Q-Sheet's internal sequencer, allowing movements to be easily recorded, reproduced, and manipulated in a sequencer-based environment.

Another controller-based automation system is that which has been incorporated into the Macintosh-based Performer Version 3.2 from Mark of the Unicorn (Fig. 11-11). This program incorporates *automated sliders* that can easily be incorporated into Performer's sequencer window for real-time continuous-controller commands. Performer allows the user to map any number of large or small sliders onto a screen, in either a vertical or horizontal configuration. For example, you might want a series of large vertical faders that could be assigned to the volume controls of a MIDI-controllable mixer and an equal number of small horizontal faders to be used for driving their associated pan controllers. Each slider may be assigned to only one channel and controller. However, they may be grouped together, allowing a master slider to correspondingly affect any number of sliders and their associated functions. Additional grouping flexibility is accomplished by the inclusion of group null points, polarity reversal, and group offsets.

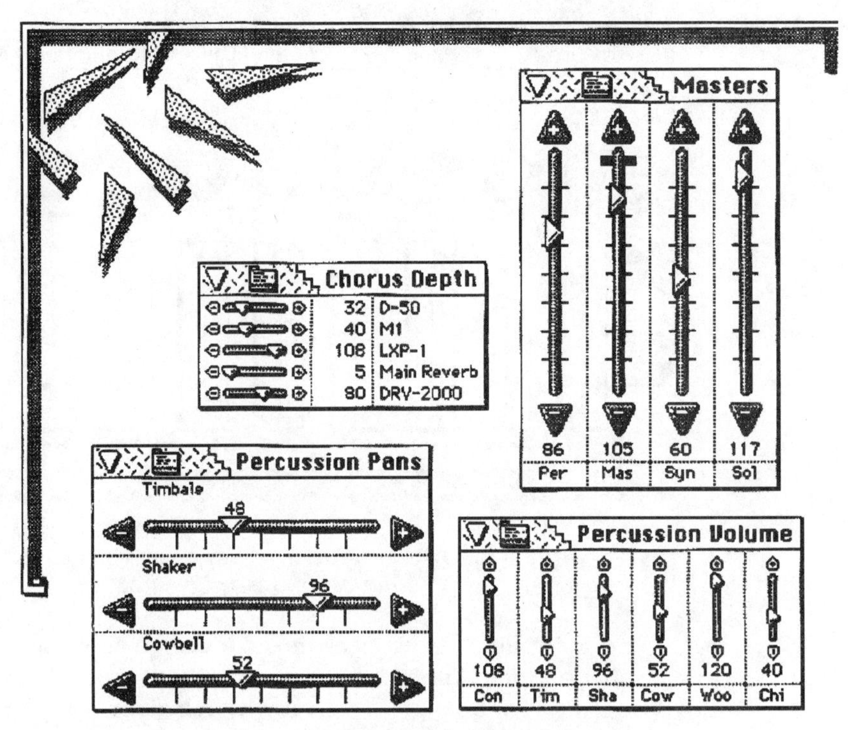

Fig. 11-11. Performer's controller-based automation
(*Courtesy of Mark of the Unicorn, Inc.*)

Console Automation

Although MIDI has not generally been implemented into larger, professional console design, many inroads are being taken towards integration into MIDI-based automation systems and smaller MIDI-controlled production mixers. Such systems are making strides toward bringing affordable and easily integrated automation packages within the grasp of the electronic musician.

MIDI-Based Automation Systems

One method of achieving low-cost automation makes use of the *VCA (voltage controlled amplifier)* to exert control over the gain and muting functions of an analog signal. VCAs are used to control program audio level through variations in a DC voltage (generally ranging from 0 to 5 volts) that is applied to the control input of the device (Fig. 11-12). As the control voltage is increased (in relation to the position of the fader), the analog signal is proportionately attenuated. Within such a dynamic automation system, these control voltage levels are encoded as a series of MIDI SysEx or controller messages and are stored as sequenced MIDI data. Upon reproduction they are used to change the audio signal levels in proportion to the way they were originally mixed.

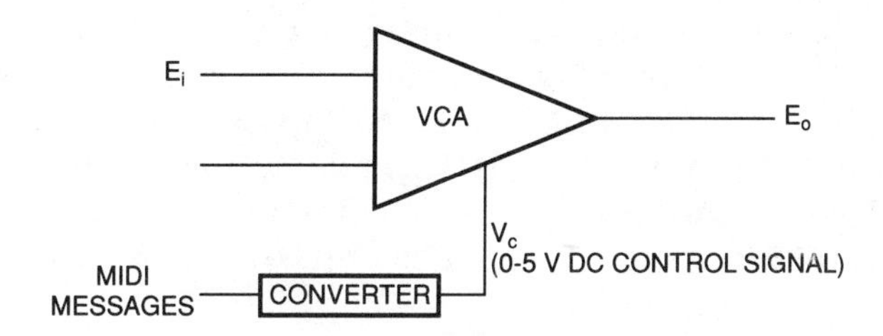

Fig. 11-12. The voltage controlled amplifier.

Devices using this technology often exist as external gain controllers which can be hooked into the insert points of an existing analog console or between the tape deck and console. This allows fader and mute functions of an otherwise nonautomated mixer or console to be memorized in real time and then reproduced under MIDI control.

The MixMate is such a self-contained 8-channel fader and mute automation system. This low-cost VCA-based interface (Fig. 11-13) allows fader movements and mute functions to be memorized in real time. Once a mix has been performed, it is a simple matter to correct any fader movement through the use of the system's update mode. After all necessary corrections have been made, MixMate will automatically merge and store these changes into its internal memory as MIDI SysEx data, which can then be recorded into a sequencer along with other music-related data.

MixMate's synchronization capabilities include the reading and generation of all forms of SMPTE and MIDI time code, in addition to MIDI Sync and SPP. Thus, this system can also be used as a synchronizer between a multitrack tape, which has been striped with SMPTE time code, and a SPP- or MTC-based sequencer.

Fig. 11-13. MixMate automation mixdown system.
(Courtesy of J.L. Cooper Electronics)

MixMate Plus software is also available. It allows a Mac or Atari ST to graphically display and control the fader, mute, disk storage, and data archiving capabilities of up to two MixMates (providing 16 channels). The device can also be externally controlled through the use of MIDI-based automation software, such as Q-Sheet A/V and Performer Version 3.

A larger-scale integrated MIDI-based automation system from J.L. Cooper Electronics is the MAGI II (Mixer Automation Gain Interface) (Fig. 11-14). Like the MixMate, this system easily interfaces with any existing mixing console to provide SMPTE-locked control over fader (level) and mute operation.

Fig. 11-14. MAGI II console automation system.
(Courtesy of J.L. Cooper Electronics)

Available in upgradable configurations of 16, 32, 48, and 64 channels, the MAGI II consists of a controller unit, rack-mounted dBx voltage-controlled amplifiers (VCAs), remote-fader unit, and system software.

The controller unit is the heart of the automation system. It is used to convert fader moves into a standard MIDI format for processing and storage within a standard Apple Macintosh (512, Plus, or SE) or Atari ST (520, 1040, or Mega) computer. These moves are locked to time code via an internal SMPTE reader generator, which reads and writes all formats. Thus, within a tape-based system (Fig. 11-15), all that is required is that the highest track number of a multitrack ATR be striped with time code. From that point on, the synchronizer will read this code and process the automated fader moves accordingly.

The MAGI II's synchronizer may be used within MIDI production to drive an external sequencer through the use of a SMPTE to MIDI time-code or direct time-lock converter, for system-wide time-code lock.

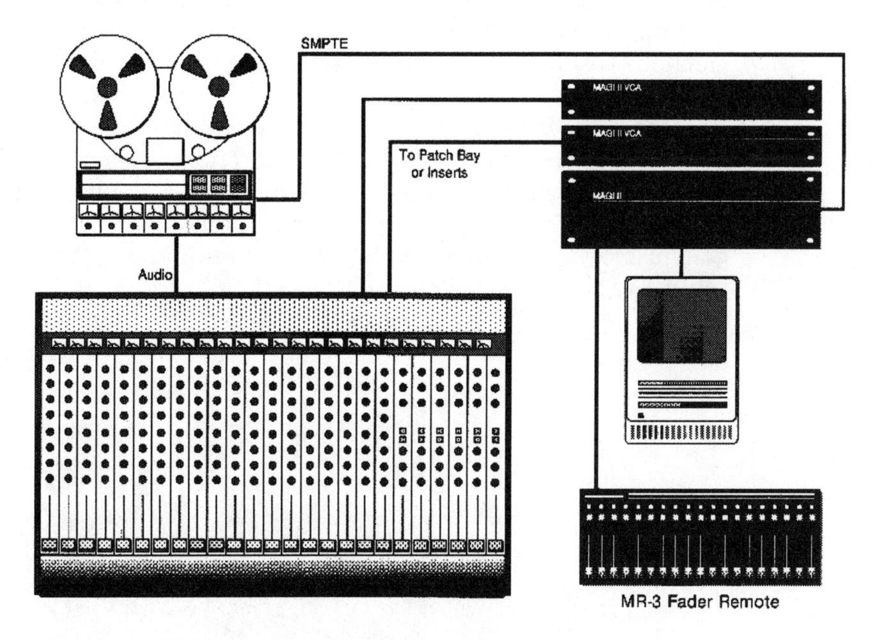

Fig. 11-15. MAGI II system configuration within a tape-based system. *(Courtesy of J.L. Cooper Electronics)*

Gain and muting functions are controlled via dBx VCAs. Each 16-channel VCA module may be placed into a mixing console, either directly into the audio path at the patch bay or channel insertion points.

Control over automation functions is accomplished by using a remote fader unit which is designed to fit on top of the console or in your lap. It is available in configurations of either 8 faders (bank switchable to 64 channels with 8 fader subgroups and 8 independent mute groups) or 20 faders offering 16 channels, and 4 groups (bank switchable to 64 channels with 4 fader subgroups and 4 independent mute subgroups).

The record, edit, and playback functions of a mix are managed through the system's disk-based software, allowing a mix to be saved on a standard 3.5-inch floppy disk. A full graphic display informs the user of fader positions, mute status, SMPTE time, fader null indication, fader subgroup and mute subgroup assignments, edit modes, and memory remaining. Additional features include MIDI event generation for automated control over MIDI-based effects devices and the ability to cut, splice, and merge multiple mixes via a cue list edit function.

MIDI-Controlled Mixers

A limited number of audio mixers which are capable of generating and responding to external MIDI SysEx or continuous-controller messages for purposes of automation are currently available. Such MIDI-controlled mixers can be either analog or digital and often vary in the degree of automation over which MIDI has control. They range from the simple MIDI control over mute functions to the creation of static snapshots for reconfiguring user-defined mixer settings, all the way to the complete dynamic control over all mixing functions via MIDI.

One such MIDI-based analog mixer is the Akai MPX820 (Fig. 11-16). This 8-in/2-out mixer is capable of storing up to 99 nondynamic snapshots in memory for control over channel level, 3-band equalization, monitor, auxiliary send, pan, and master. The fade times between two different snapshot settings is programmable from 40 milliseconds to 30 seconds, allowing for fades, pans, and EQ settings to be automated over time.

Fig. 11-16. Akai MPX820 mixer. *(Courtesy of Akai Professional)*

The Yamaha DMP-7 is a MIDI-based mixer (Fig. 11-17) which employs a fully digital 8-in/2-out digital mixing format. Each input channel includes a 16-bit linear A/D converter (sampling at 44.1 kHz), a 3-band digital parametric equalizer, panning, effects, and motorized channel faders. Effects buses 1 and 2 provide 15 different digital effects via two internal digital signal processors, while effects bus 3 offers five digital effects, an external effects send, and stereo effects returns. The output buses feature digital compression, externally controllable output levels, D/A converters, and both balanced and unbalanced outputs. Using a digital bus-cascade feature, up to 4 DMP-7s can be cascaded to create a 32 x 2 mixing format with 12 DSP multi-effects systems.

Fig. 11-17. Yamaha DMP-7 digital mixer.
(*Courtesy of the Yamaha Corporation*)

Up to 32 mixing and processing settings may be internally stored as snapshots within the DMP-7, while another 67 may be saved onto an external RAM cartridge. In addition, all dynamic signal functions may be controlled in real time via MIDI through the use of a sequencer or specially designed software package.

As a digital I/O version of the DMP-7 mixer, the DMP-7 has been designed to offer a variety of standard digital input/output sources including Yamaha DSP-LSI, Sony PCM-3324, Mitsubishi X-850, AES/EBU standard, and CD/DAT. DMP-7D users requiring analog inputs can use the optional AD808 A/D conversion box which includes eight analog XLR inputs and an RS-422 digital output port for connection to the DMP-7D.

The Yamaha DMP-11 digital mixing processor (Fig. 11-18) is a rack-mounted, economical version of the DMP-7. This device integrates an 8-in/2-out fully-digital line level mixer with two internal digital effects systems.

Fig. 11-18. Yamaha DMP-11 digital mixer.
(*Courtesy of the Yamaha Corporation*)

Unlike the DMP-7, the DMP-11 does not incorporate motorized faders. Instead, level displays of these fully automated controls can be called up on the LCD panel to graphically show memorized fader settings.

This device features 96 internal memories which are able to store all console parameters in a snapshot configuration. These may be instantly recalled either from the front panel or through the use of MIDI program change messages. Additionally, the DMP-11 can be directly connected to a MIDI sequencer, allowing real-time recording and editing of mix and effects operations, or remotely operated via a master controller or MIDI controller software.

The Dedicated Mixing Data Line

Many newer sequencers for the Macintosh, IBM, and Atari are capable of simultaneously addressing more than one MIDI data line. When this option is available to you, it is a wise idea to assign a dedicated MIDI line for performing the task of transmitting MIDI mixing data (Fig. 11-19). This will reduce the possibility of MIDI delays and data clogs should large amounts of controller messages be generated by a hardware- or software-based MIDI controller and MIDI controlled mixer.

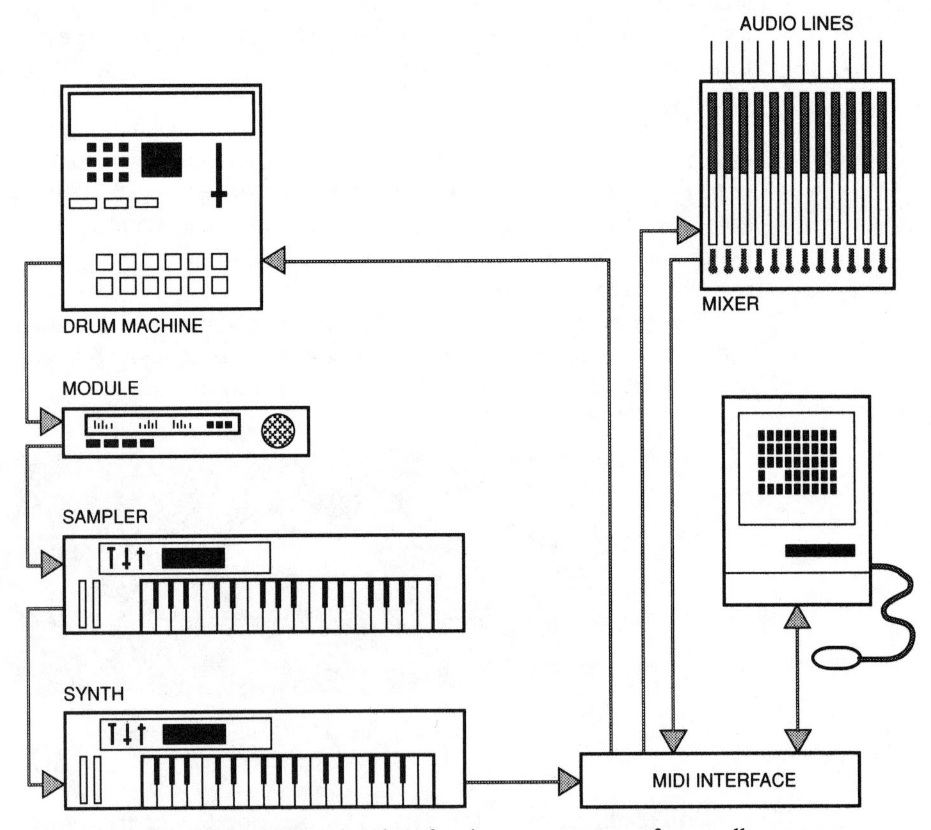

Fig. 11-19. Isolated MIDI data line for the transmission of controller messages during mixing.

Appendix A

The MIDI 1.0 Specification*

Introduction

MIDI is the acronym for Musical Instrument Digital Interface.

MIDI enables synthesizers, sequencers, home computers, rhythm machines, etc. to be interconnected through a standard interface.

Each MIDI-equipped instrument usually contains a receiver and a transmitter. Some instruments may contain only a receiver or transmitter. The receiver receives messages in MIDI format and executes MIDI commands. It consists of an opto-isolator, Universal Asynchronous Receiver/Transmitter (UART), and other hardware needed to perform the intended functions. The transmitter originates messages in MIDI format, and transmits them by way of a UART and line driver.

The MIDI standard hardware and data format are defined in this specification.

Conventions

Status and Data bytes given in Tables I through VI are given in binary.

Numbers followed by an "H" are in hexadecimal.

All other numbers are in decimal.

---

*MIDI Manufacturers Association, 5316 West 57th Street, Los Angeles, CA, 90056

Hardware

The interface operates at 31.25 (± 1%) Kbaud, asynchronous, with a start bit, 8 data bits (D0 to D7), and a stop bit. This makes a total of 10 bits for a period of 320 microseconds per serial byte.

Circuit: 5 mA current loop type. Logical 0 is current ON. One output shall drive one and only one input. The receiver shall be opto-isolated and require less than 5 mA to turn on. Sharp PC-900 and HP 6N138 opto-isolators have been found acceptable. Other high-speed opto-isolators may be satisfactory. Rise and fall times should be less than 2 microseconds.

Connectors: DIN 5 pin (180 degree) female panel mount receptacle. An example is the SWITCHCRAFT 57 GB5F. The connectors shall be labelled "MIDI IN" and "MIDI OUT." Note that pins 1 and 3 are not used, and should be left unconnected in the receiver and transmitter.

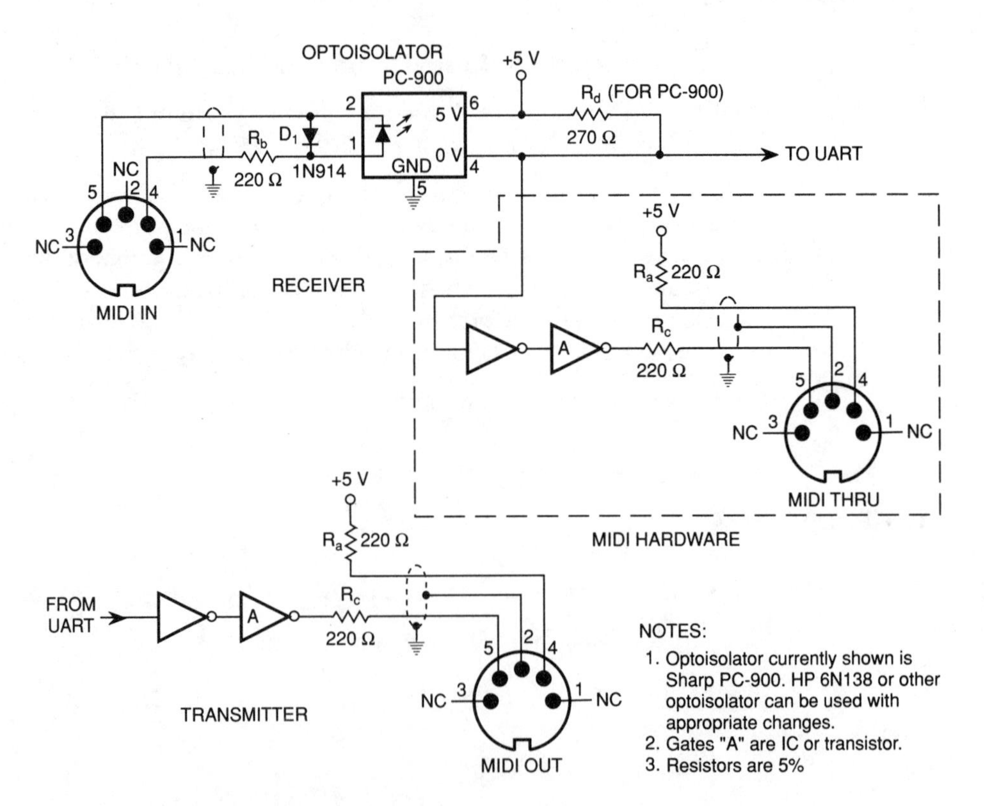

Fig. A-1. Standard hardware configuration for MIDI in, out, and thru ports.

Cables shall have a maximum length of fifty feet (15 meters), and shall be terminated on each end by a corresponding 5-pin DIN male plug, such as the SWITCHCRAFT 05GM5M. The cable shall be shielded twisted pair, with the shield connected to pin 2 at both ends.

A "MIDI THRU" output may be provided if needed, which provides a direct copy of data coming in MIDI IN. For very long chain lengths (more than three instruments), higher-speed opto-isolators must be used to avoid additive rise/fall time errors which affect pulse width duty cycle.

Data Format

All MIDI communication is achieved through multi-byte "messages" consisting of one Status byte followed by one or two Data bytes, except Real-Time and Exclusive messages (see below).

Message Types

Messages are divided into two main categories: Channel and System.

Channel

Channel messages contain a four-bit number in the Status byte which address the message specifically to one of sixteen channels. These messages are thereby intended for any units in a system whose channel number matches the channel number encoded into the Status byte.

There are two types of Channel messages: Voice and Mode.

Voice–To control the instrument's voices, Voice messages are sent over the Voice Channels.

Mode–To define the instrument's response to Voice messages, Mode messages are sent over the instrument's Basic Channel.

System

System messages are not encoded with channel numbers.

There are three types of System messages: Common, Real-Time, and Exclusive.

Common–Common messages are intended for all units in a system.

Real-Time–Real-Time messages are intended for all units in a system.

They contain Status bytes only—no Data bytes. Real-Time messages may be sent at any time—even between bytes of a message which has a different status. In such cases the Real-Time message is either ignored or acted upon, after which the receiving process resumes under the previous status.

Exclusive–Exclusive messages can contain any number of Data bytes, and are terminated by an End of Exclusive (EOX) or any other Status byte. These messages include a Manufacturer's Identification (ID) code. If the receiver does not recognize the ID code, it should ignore the ensuing data.

So that other users can fully access MIDI instruments, manufacturers should publish the format of data following their ID code. Only the manufacturer can update the format following their ID.

Data Types

Status Bytes

Status bytes are eight-bit binary numbers in which the Most Significant Bit (MSB) is set (binary 1). Status bytes serve to identify the message type; that is, the purpose of the Data bytes which follow the Status byte.

Except for Real-Time messages, new Status bytes will always command the receiver to adopt their status, even if the new Status is received before the last message was completed.

Running Status—For Voice and Mode messages only, when a Status byte is received and processed, the receiver will remain in that status until a different Status byte is received. Therefore, if the same Status byte would be repeated, it may (optionally) be omitted so that only the correct number of Data bytes need be sent. Under Running Status, then, a complete message need only consist of specified Data bytes sent in the specified order.

The Running Status feature is especially useful for communicating long strings of Note On/Off messages, where "Note On with Velocity of 0" is used for Note Off. (A separate Note Off Status byte is also available.)

Running Status will be stopped when any other Status byte intervenes, except that Real-Time messages will only interrupt the Running Status temporarily.

Unimplemented Status—Any status bytes received for functions which the receiver has not implemented should be ignored, and subsequent data bytes ignored.

Undefined Status—Undefined Status bytes must not be used. Care should be taken to prevent illegal messages from being sent during power-up or power-down. If undefined Status bytes are received, they should be ignored, as should subsequent Data bytes.

Data Bytes

Following the Status byte, there are (except for Real-Time messages) one or two Data bytes which carry the content of the message. Data bytes are eight-bit binary numbers in which the MSB is reset (binary 0). The number and range of Data bytes which must follow each Status byte are specified in the tables which follow. For each Status byte the correct number of Data bytes must always be sent. Inside the receiver, action on the message should wait until all Data bytes required under the current status are received. Receivers should ignore Data bytes which have not been properly preceded by a valid Status byte (with the exception of "Running Status," above).

Channel Modes

Synthesizers contain sound generation elements called voices. Voice assignment is the algorithmic process of routing Note On/Off data from the keyboard to the voices so that the musical notes are correctly played with accurate timing.

When MIDI is implemented, the relationship between the sixteen available MIDI channels and the synthesizer's voice assignment must be defined. Several Mode messages are available for this purpose (see Table III). They are Omni (On/Off), Poly, and Mono. Poly and Mono are mutually exclusive, i.e., Poly Select disables Mono, and vice versa. Omni, when on, enables the receiver to receive Voice messages in all Voice Channels without discrimination. When Omni is off, the receiver will accept Voice messages from only the selected Voice Channel(s). Mono, when on, restricts the assignment of Voices to just one voice per Voice Channel (Monophonic.) When Mono is off (=Poly On), any number of voices may be allocated by the Receiver's normal voice assignment algorithm (Polyphonic).

For a receiver assigned to Basic Channel "N," the four possible modes arising from the two Mode messages are:

Mode	Omni		
1	On	Poly	Voice messages are received from all Voice Channels and assigned to voices polyphonically.
2	On	Mono	Voice messages are received from all Voice Channels, and control only one voice, monophonically.
3	Off	Poly	Voice messages are received in Voice Channel N only, and are assigned to voices polyphonically.
4	Off	Mono	Voice messages are received in Voice Channels N thru N+M−1, and assigned monophonically to voices 1 thru M, respectively. The number of voices M is specified by the third byte of the Mono Mode Message.

Four modes are applied to transmitters (also assigned to Basic Channel N). Transmitters with no channel selection capability will normally transmit on Basic Channel 1 (N=0).

Mode	Omni		
1	On	Poly	All voice messages are transmitted in Channel N.
2	On	Mono	Voice messages for one voice are sent in Channel N.
3	Off	Poly	Voice messages for all voices are sent in Channel N.
4	Off	Mono	Voice messages for voices 1 thru M are transmitted in Voice Channels N thru N+M−1, respectively. (Single voice per channel.)

A MIDI receiver or transmitter can operate under one and only one mode at a time. Usually the receiver and transmitter will be in the same mode. If a mode cannot be honored by the receiver, it may ignore the message (and any subsequent data bytes), or it may switch to an alternate mode (usually Mode 1, Omni On/Poly).

Mode messages will be recognized by a receiver only when sent in the Basic Channel to which the receiver has been assigned, regardless of the current mode. Voice messages may be received in the Basic Channel and in other channels (which are all called Voice Channels), which are related specifically to the Basic Channel by the rules above, depending on which mode has been selected.

A MIDI receiver may be assigned to one or more Basic Channels by default or by user control. For example, an eight-voice synthesizer might be assigned to Basic Channel 1 on power-up. The user could then switch the instrument to be configured as two four-voice synthesizers, each assigned to its own Basic Channel. Separate Mode messages would then be sent to each four-voice synthesizer, just as if they were physically separate instruments.

Power-Up Default Conditions

On power-up all instruments should default to Mode #1. Except for Note On/Off Status, all Voice messages should be disabled. Spurious or undefined transmissions must be suppressed.

Table I. Summary of Status Bytes

Status D7—D0	# of Data Bytes	Description
		Channel Voice Messages
1000nnnn	2	Note Off event
1001nnnn	2	Note On event (velocity=0: Note Off)
1010nnnn	2	Polyphonic key pressure/after touch
1011nnnn	2	Control change
1100nnnn	1	Program change
1101nnnn	1	Channel pressure/after touch
1110nnnn	2	Pitch bend change
		Channel Mode Messages
1011nnnn	2	Selects Channel Mode
		System Messages
11110000	*****	System Exclusive
11110sss	0 to 2	System Common
11111ttt	0	System Real Time

Notes:

nnnn: N-1, where N = Channel #,
i.e., 0000 is Channel 1.
0001 is Channel 2.
.
.
.
1111 is Channel 16.

*****: 0iiiiiii, data, ..., EOX

iiiiiii: Identification

sss: 1 to 7

ttt: 0 to 7

Table II. Channel Voice Messages

Status	Data Bytes	Description
1000nnnn	0kkkkkkk	Note Off (see notes 1-4)
	0vvvvvvv	vvvvvvv: note off velocity
1001nnnn	0kkkkkkk	Note On (see notes 1-4)
	0vvvvvvv	vvvvvvv - 0: velocity
		vvvvvvv = 0: note off
1010nnnn	0kkkkkkk	Polyphonic Key Pressure (After-Touch)
	0vvvvvvv	vvvvvvv: pressure value
1011nnnn	0ccccccc	Control Change
	0vvvvvvv	ccccccc: control # (0-121)
		(see notes 5-8)
		vvvvvvv: control value
		ccccccc = 122 thru 127: Reserved.
		(See Table III)
1100nnnn	0ppppppp	Program Change
		ppppppp: program number (0-127)
1101nnnn	0vvvvvvv	Channel Pressure (After-Touch)
		vvvvvvv: pressure value
1110nnnn	0vvvvvvv	Pitch Bend Change LSB (see note 10)
	0vvvvvvv	Pitch Bend Change MSB

Notes:

1. nnnn: Voice Channel # (1-16, coded as defined in Table I notes)

2. kkkkkkk: note # (0 - 127)
   kkkkkkk = 60: Middle C of keyboard
   0 12 24 36 48 60 72 84 96 108 120 127

   ac c c c c c c c
   |——— piano range ———|

3. vvvvvvv: key velocity
   A logarithmic scale would be advisable.
   0 1 64 127

   off ppp pp p mp mf f fff fff

   vvvvvvv = 64: in case of no velocity sensors
   vvvvvvv = 0: Note Off, with velocity = 64

4. Any Note On message sent should be balanced by sending a Note Off message for that note in that channel at some later time.

5. ccccccc: control number

ccccccc	Description
0	Continuous Controller 0 MSB
1	Continuous Controller 1 MSB (MODULATION BENDER)
2	Continuous Controller 2 MSB
3	Continuous Controller 3 MSB
4-31	Continuous Controllers 4-31 MSB
32	Continuous Controller 0 LSB
33	Continuous Controller 1 LSB (MODULATION BENDER)
34	Continuous Controller 2 LSB
35	Continuous Controller 3 LSB
36-63	Continuous Controllers 4-31 LSB
64-95	Switches (On/Off)
96-121	Undefined
122-127	Reserved for Channel Mode messages (see Table III).

6. All controllers are specifically defined by agreement of the MIDI Manu facturers Association (MMA) and the Japan MIDI Standards Committee (JMSC). Manufacturers can request through the MMA or JMSC that logical controllers be assigned to physical ones as necessary. The controller allocation table must be provided in the user's operation manual.

7. Continuous controllers are divided into Most Significant and Least Signifi cant Bytes. If only seven bits of resolution are needed for any particular controllers, only the MSB is sent. It is not necessary to send the LSB. If more resolution is needed, then both are sent, first the MSB, then the LSB. If only the LSB has changed in value, the LSB may be sent without re-sending the MSB.

8. vvvvvvv: control value (MSB)
   (for controllers)

   0————————————————127
   min max

   (for switches)

   0 - - - - - - - - - - 127
   off on

   Numbers 1 through 126, inclusive, are ignored.

9. Any messages (e.g., Note On), which are sent successively under the same status, can be sent without a Status byte until a different Status byte is needed.

10. Sensitivity of the pitch bender is selected in the receiver. Center position value (no pitch change) is 2000H, which would be transmitted EnH-00H-40H.

Table III. Channel Mode Messages

Status	Data Bytes	Description
1011nnnn	0ccccccc 0vvvvvvv	Mode Messages ccccccc = 122: Local Control vvvvvvv = 0, Local Control Off vvvvvvv = 127, Local Control On ccccccc = 123: All Notes Off vvvvvvv = 0 ccccccc = 124: Omni Mode Off (All Notes Off) vvvvvvv = 0 ccccccc = 125: Omni Mode On (All Notes Off) vvvvvvv = 0 ccccccc = 126: Mono Mode On (Poly Mode Off) (All Notes Off) vvvvvvv = M, where M is the number of channels. vvvvvvv = 0, the number of channels equals the number of voices in the receiver. ccccccc = 127: Poly Mode On (Mono Mode Off) vvvvvvv = 0 (All Notes Off)

Notes:

1. nnnn: Basic Channel # (1-16, coded as defined in Table I)

2. Messages 123 thru 127 function as All Notes Off messages. They will turn off all voices controlled by the assigned Basic Channel. Except for message 123, All Notes Off, they should not be sent periodically, but only for a specific purpose. In no case should they be used in lieu of Note Off commands to turn off notes which have been previously turned on. Therefore any All Notes Off command (123-127) may be ignored by receiver with no possibility of notes staying on, since any Note On command must have a corresponding specific Note Off command.

3. Control Change #122, Local Control, is optionally used to interrupt the internal control path between the keyboard, for example, and the sound-generating circuitry. If 0 (Local Off message) is received, the path is disconnected: the keyboard data goes only to MIDI and the sound-generating circuitry is controlled only by incoming MIDI data. If a 7FH (Local On message) is received, normal operation is restored.

4. The third byte of "Mono" specifies the number of channels in which Monophonic Voice messages are to be sent. This number, "M", is a number between 1 and 16. The channel(s) being used, then, will be the current

Basic Channel (=N) thru N+M−1 up to a maximum of 16. If M=0, this is a special case directing the receiver to assign all its voices, one per channel, from the Basic Channel N through 16.

Table IV. System Common Messages

Status	Data Bytes	Description
11110001		Undefined
11110010		Song Position Pointer
	0lllllll	lllllll: (Least significant)
	0hhhhhhh	hhhhhhh: (Most significant)
11110011	0sssssss	Song Select
		sssssss: Song #
11110100		Undefined
11110101		Undefined
11110110	none	Tune Request
11110111	none	EOX: "End of System Exclusive" flag

1. Song Position Pointer: Is an internal register which holds the number of MIDI beats (1 beat = 6 MIDI clocks) since the start of the song. Normally it is set to 0 when the START switch is pressed, which starts sequence playback. It then increments with every sixth MIDI clock receipt, until STOP is pressed. If CONTINUE is pressed, it continues to increment. It can be arbitrarily preset (to a resolution of 1 beat) by the SONG POSITION POINTER message.
2. Song Select: Specifies which song or sequence is to be played upon receipt of a Start (Real-Time) message.
3. Tune Request: Used with analog synthesizers to request them to tune their oscillators.
4. EOX: Used as a flag to indicate the end of a System Exclusive transmission (see Table VI).

Table V. System Real Time Messages

Status	Data Bytes	Description
11111000		Timing Clock
11111001		Undefined
11111010		Start
11111011		Continue
11111100		Stop
11111101		Undefined
11111110		Active Sensing
11111111		System Reset

Notes:

1. The System Real Time messages are for synchronizing all of the system in real time.
2. The System Real Time messages can be sent at any time. Any messages which consist of two or more bytes may be split to insert Real Time messages.
3. Timing clock (F8H)
   The system is synchronized with this clock, which is sent at a rate of 24 clocks/quarter note.
4. Start (from the beginning of song) (FAH)
   This byte is immediately sent when the PLAY switch on the master (e.g., sequencer or rhythm unit) is pressed.
5. Continue (FBH)
   This is sent when the CONTINUE switch is hit. A sequence will continue at the time of the next clock.
6. Stop (FCH)
   This byte is immediately sent when the STOP switch is hit. It will stop the sequence.
7. Active Sensing (FEH)
   Use of this message is optional, for either receivers or transmitters. This is a "dummy" Status byte that is sent every 300 ms (max), whenever there is no other activity on MIDI. The receiver will operate normally if it never receives FEH. Otherwise, if FEH is ever received, the receiver will expect to receive FEH or a transmission of any type every 300 ms (max). If a period of 300 ms passes with no activity, the receiver will turn off the voices and return to normal operation.
8. System Reset (FFH)
   This message initializes all of the system to the condition of just having turned on power. The system Reset message should be used sparingly, preferably under manual command only. In particular, it should not be sent automatically on power up.

Table VI. System Exclusive Messages

Status	Data Bytes	Description
11110000		Bulk dump etc.
	0iiiiiii	iiiiiii: identification
	•	
	(0*******)	
	•	Any number of bytes may be sent here, for any purpose, as long as they all have a zero in the most significant bit.
	(0*******)	
	•	
	11110111	EOX: "End of System Exclusive"

Notes:

1. iiiiiii: identification ID (0-127)
2. All bytes between the System Exclusive Status byte and EOX or the next Status byte must have zeroes in the MSB.
3. The ID number can be obtained from the MMA or JMSC.
4. In no case should other Status or Data bytes (except Real-Time) be interleaved with System Exclusive, regardless of whether or not the ID code

Appendix B

The MIDI Implementation Chart

It is not always necessary for a MIDI device to transmit or receive every type of MIDI message that is defined by the MIDI specification. Certain messages may not relate to the function of a device. For example, it is only required of a synth module that it respond to note-on/off messages since there is no built-in keyboard controller for transmitting them. Other devices might limit various MIDI messages due to such factors as design limitations or cost-effectiveness. For instance, should a velocity-sensitive keyboard controller be used to control a synthesizer that does not respond to velocity messages, no amount of keyboard banging will result in a change in volume. As a result, the velocity message is simply ignored by the synthesizer.

To insure that two or more MIDI devices will be able to communicate MIDI events effectively, the MMA and the JMSC have devised the *MIDI Implementation Chart* (Fig. B-1), which relates all of the MIDI capabilities of a specific MIDI device to the user at a glance using a standardized printed format.

From the user's standpoint, when considering a new piece of equipment, it is always wise to compare its implementation chart with other devices within the existing MIDI system. This will ensure that the device will recognize existing messages and/or add to the capabilities of the current system.

Guidelines for Using the Chart

The MMA specifies that the MIDI implementation chart be printed the same size using a standardized spreadsheet format consisting of 4 columns by 12 rows. The first column lists the MIDI function in question. The second lists information relating to whether (or how) the device transmits this function's data. The third lists whether (or how) the device recognizes (receives) this data, and the final column is used for additional remarks by the manufacturer.

Function ···		Transmitted	Recognized	Remarks
Basic Channel	Default	1 – 16	1 – 16	memorized
	Changed	1 – 16	1 – 16	
Mode	Default	Mode 3, 4	Mode 3	memorized
	Messages	OMNI OFF, MONO POLY	×	
	Alterd	*********		
Note Number		0 – 127	0 – 127	
	True Voice	*********	12 – 108	
Velocity	Note ON	○ v = 1 – 127	○ v = 1 – 127	
	Note OFF	× 9n v = 0	×	
After Touch	Key's	×	×	
	Ch's	×	×	
Pitch Bender		○	○ 0 – 24 semitone	
Control Change	1	○	○	Modulation
	2 – 5	×	×	
	6	**	**	Data Entry MSB
	7	○	○	Volume
	8 – 15	×	×	
	16	×	○	General Purpose Control-1
	17 – 37	×	×	
	38	**	×	Data Entry LSB
	39 – 63	×	×	
	64	×	○	Hold 1
	65 – 80	×	×	
	81	×	○	General Purpose Control-1
	82 – 99	×	×	
	100 – 101	** (0)	** (0)	RPC LSB, MSB
	102 – 120	×	×	
	121	○	○	Reset All Controllers
Prog Change		○ 0 – 127	○ 0 – 127	
	True #	*********	0 – 127	
System Exclusive		○	○	
System Common	Song Pos	×	×	
	Song Sel	×	×	
	Tune	×	×	
System Real Time	Clock	×	×	
	Commands	×	×	
Aux Message	Local ON/OFF	×	×	
	All Notes OFF	×	○	
	Active Sense	○	○	
	Reset	×	×	
Notes		*Control Change messages from 0 to 95 which are recognized through Control channel are transmitted throgh all the channels which are used in Branches. However, General Purpose Control -1 and General Purpose Control – 6 are converted into the same functions as the FC-100 EV-5 assign and the FC-100 Switch assign in the System Setup, and are transmitted. **RPC = Registered Parameter Control Number RPC # 0 : Bender Range The value of parameter is to be determined by entering data.		

Mode 1 : OMNI ON, POLY Mode 2 : OMNI ON, MONO ○ : Yes
Mode 3 : OMNI OFF, POLY Mode 4 : OMNI OFF, MONO × : No

Fig. B-1. Example of a MIDI Implementation Chart.

Despite efforts at standardization, slight inconsistencies within the chart's specifications allow for variations in the symbols, abbreviations, spelling, etc. that can be used by different manufacturers. The following guidelines provide a basic understanding of these differences.

- In general, the symbol O is used to indicate that a MIDI function *is* implemented, while an X is used to show that the function is *not* implemented. However, some charts may use an X to equal a *yes* and an O to equal a *no*. This will usually be indicated within a key at the lower right-hand corner of the chart.
- OX or "*" is used to indicate a selectable function. Further information on the range or type of selectability will be placed within the remarks column.
- MIDI modes are listed as follows:

  Mode 1 (omni on, poly)
  Mode 2 (omni on, mono)
  Mode 3 (omni off, poly)
  Mode 4 (omni off, mono)

  These will often be listed at the bottom of the chart. Occasionally abbreviations of these modes (i.e., omni on/off, omni on or poly) may be used by a manufacturer.

Detailed Explanation of the Chart

The following is a detailed explanation of the various functions and their related categories that are found within the chart.

Header

The *header* provides the user with the model number, brief description, date, and version number of the device.

Basic Channel

Basic channel indicates which MIDI channels are used by the device to transmit and receive data. The subheadings for this function are *default* and *changed*.

- *Default*: This indicates which MIDI channel is in use when the device is first turned on.
- *Changed*: This indicates which of the MIDI channels can be addressed after the device is first turned on.

Mode

Mode indicates which of the MIDI modes may be used by the device. The subheadings for this function are default, messages, and altered.

- *Default*: This indicates which of the four MIDI modes is active when the device is first turned on.
- *Messages*: This describes which of the four MIDI modes can be transmitted or recognized by the device.
- *Altered*: This refers to mode messages which cannot be recognized by the device. It may be followed by a description of the mode that the device automatically enters into upon receiving a request message for an unavailable mode.

Note Number

The transmitted *note number* indicates the range of MIDI note numbers that are transmitted by a device. The maximum possible range spans from 0–127, while 21–108 corresponds to the 88 keys of an extended keyboard controller. Should the note number be greater than the actual number of keys on a keyboard device, a key transposition feature is indicated.

The recognized note number indicates the range of MIDI note numbers that can be recognized by a device. MIDI notes that are out of this range shall be ignored by this device. A second note number range, known as *true voice*, indicates the number of notes the device can actually play. Recognized notes that are out of the actual voice range are transposed up or down in octaves until they fall within this range.

Velocity

This category indicates whether the device is capable of transmitting or receiving attack- and release-velocity messages. The subheadings for this function are *note on* and *note off*.

- *Note on*: This indicates if the device is capable of transmitting and responding to variable-velocity (attack) messages. Not all dynamically controllable devices respond to the full velocity range (1–127). Some devices, such as drum machines, respond to a finite number of velocity steps.

- *Note off*: Indicates whether the device is capable of transmitting and responding to variable release velocity messages. Many devices use a message (note-on velocity = 0) to indicate a note-off condition. This is often indicated in the chart by *9NH v=0* or *$9n 00*, which is the hexadecimal equivalent for this message.

After Touch

After touch indicates how pressure data is transmitted or received. The subheadings for this function are *key's* and *ch's*.

- *Key's*: This indicates if the device will transmit or receive independent polyphonic-pressure messages for each key.
- *Ch's*: Indicates whether the device is capable of transmitting or receiving channel-pressure changes (a common after-touch mode, providing one pressure value for an entire MIDI channel).

Pitch Bender

Pitch bender indicates if the device is capable of transmitting or receiving pitch-bend information. If so, the remarks column will often give information as to the pitch bend range and resolution.

Control Change

Control change indicates whether the device is capable of transmitting or receiving continuous-controller messages. The chart will often list which of these messages are supported in addition to providing a detailed breakdown of their parameters within the remarks column.

Program Change

This category indicates if the device is capable of transmitting or receiving *program-change messages*. *True #* indicates the message numbers that are actually supported by the device's program-change buttons.

System Exclusive

This indicates if the device is capable of transmitting and receiving *system exclusive data*. The remarks column will often give general information as to which type of SysEx data is supported. However, more detailed data will generally be provided within the device's manual.

System Common

This indicates whether the device is capable of transmitting or receiving the different types of *system common messages*, such as SPP, MIDI time code, song select, and tune-request messages.

System Real Time

This category indicates whether the device can transmit or receive *system real-time messages*. The subheadings for this function are *clock* and *commands*.

- *Clock*: This refers to the device's ability to receive or transmit MIDI clock messages. A device that can transmit MIDI clock may be used to provide master timing information within a MIDI system, while a device capable of receiving clock data may only be slaved to other MIDI devices.
- *Commands*: This indicates whether the device is capable of transmitting or responding to start, stop, and continue messages.

Aux Messages

This indicates if a device is capable of transmitting or receiving local control-on/off, all notes-off, active-sensing, and system-reset messages.

Notes

This area is used by the manufacturer to comment on any function or implementation particular to the specific MIDI device.

Appendix C

Continued Education

As the equipment which is facing the electronic musician places an ever-increasing emphasis upon technology, education must play a greater role in the understanding of basic industry skills. Education can take many forms, ranging from a formal education to simply keeping abreast of industry directions from the many industry magazines.

The following resource listings are recommended for anyone wishing to further their education in electronic music technology.

Magazines

Electronic Musician
P.O. Box 41094
Nashville, TN 37204
(800)888-5139

Keyboard
P.O. Box 58528
Boulder, CO 80322
(800)289-9919
(303)447-9330 (in Colorado)

International Musician & Recording World
Stonehart Subcription Services
Hainault Road, Little Heath
Romford, Essex, England RM6-5NP
01-597-7335

International MIDI Association Bulletin
5316 West 57th Street
Los Angeles, CA 90056
(213)649-6434
(213)215-3380 (Fax)

Home & Studio Recording (US)
Music Maker Publications, Inc.
22024 Lassen Street, Suite 118
Chatsworth, CA 91311
(818)407-0744

Rhythm
Music Maker Publications, Inc.
22024 Lassen Street, Suite 118
Chatsworth, CA 91311
(818)407-0744

Modern Drummer
P.O. Box 480
Mount Morris, IL 61054
(800)435-0715

Musician
Box 1923
Marion, OH 43305
(800)347-6969
(614)382-3322

Computer Music Journal
Journals Department
The MIT Press
55 Hayward Street
Cambridge, MA 02142

Books

Many of the following publications are available from *The Mix Bookshelf Catalog*, 6400 Hollis Street, Suite #12, Emeryville, CA 94608, (800)233-9604 or (415)653-3307.

MIDI for Musicians, Craig Anderton, Amsco Publications, New York, 1986.

How MIDI Works, Dan Walker, Alexander Publishing, Newbury Park, CA, 1989.

Using MIDI, Helen Casabona and David Frederick, GPI Publications, Cupertino, CA, 1987.

MIDI Home Studio, Howard Massey, Music Sales Corporation, New York, NY, 1988.

MIDI Programmer's Handbook, Steve De Furia and Joe Scacciaferro, M&T Publishing, Inc., Redwood City, CA, 1989.

MIDI Programming for the Macintosh (with disk), Steve De Furia and Joe Scacciaferro, M&T Publishing, Inc., Redwood, CA, 1988.

Music Through MIDI, Michael Boom, Microsoft Press, Redmond, WA, 1987.

Syncronization: From Reel to Reel, Jeffrey Rona, Hal Leonard Corporation, Milwaukee, WI, 1990.

Music and the Macintosh, Geary Yelton, MIDI America, Atlanta, GA, 1989

MIDI: A Comprehensive Introduction, Joseph Rothstein, A-R Editions, Inc., Madison, WI, 1991.

MIDI Systems and Control, Francis Rumsey, Focal Press, Boston, MA, 1990.

Organizations

International MIDI Association
5316 West 57th Street
Los Angeles, CA 90056
(213)649-MIDI
(213)215-3380 (Fax)

MIDI Manufacturers Association
5316 West 57th Street
Los Angeles, CA 90056
(213)649-MIDI

Apple Programmers and Developers Association (APDA)
Apple Computers, Inc.
20525 Mariani Avenue, M/S 33G
Cupertino, CA 95014
(800)282-APDA (US)
(800)637-0029 (Canada)
(408)562-3910 (International)

Networks

Computer bulletin board networks are available that allow the user to access large databases of information and to communicate with other electronic musicians via computer/phone modem links. These services allow the user to download and upload music and computer-related programs, patch data, computer utilities and text data (often information and reviews relating to new equipment). Users may also commonly communicate using public and private forums (allowing the users to voice their comments and ideas to other users and manufacturers) in addition to electronic mail (E-mail) services.

These bulletin board services are available to any user equipped with any popular model of personal computer and modem communications device. It is usually required that the user pay an initial sign-up fee, after which the user is billed at a specified user's rate for on-line usage. Often software manufacturers will provide a waiver of the initial sign-up fee, enabling the user to get on-line with a network that supports the software and/or manufacturer in question.

PAN (Performing Artists Network)
P.O. Box 162
Skippack, PA 19474
(215)584-0300

COMPUSERVE—MIDI/Music Forum
P.O. Box 20212
Columbus, OH 43220
(800)848-8199

MIDI-Exchange BBS
P.O. Box 640608
San Francisco, CA 94164-0608
(415)771-1788 (300 or 1200 baud)

Glossary

Auxiliary controllers: These are external controlling devices used in conjunction with a main instrument or controller. Some examples of such controllers are foot pedals, the switch pedal (i.e., sustain), the breath controller, the pitch-bending wheel, and the modulation wheel.

Bits: Electronic data that is used to represent the binary numeric system of 0s and 1s, or on and off.

Bytes: A grouping of eight bits into a digital word.

Central processing unit (CPU): A microprocessor or computer which is used to perform complex task-related functions. Within an electronic musical instrument, it is a small, dedicated computer system for handling the many performance- and control-related messages and commands that must be processed in real time.

Channel messages: These are messages that are assigned to a specific MIDI channel within a system or device.

Channel-voice messages: These are used to transmit real-time performance data throughout a connected MIDI system. They are generated whenever the controller of a MIDI instrument is played, selected, or varied by the performer.

Click sync/click track: This refers to the metronomic audio clicks that are generated by electronic devices to communicate tempo.

Cue-list/list-edit sequencer: A sequencing program that places data which relates to a specific cue or range of cues within a vertical, sequential-edit list.

Daisy chain: This method for distributing data within a MIDI system is used to distribute MIDI data to every device within a system by transmitting data to the first device and subsequently passing an exact copy of this data through to each device within the chain.

Data bytes: Contains a value (generally 0–127) that is to be attached to the status byte.

Direct time lock (DTL) and enhanced direct time lock (DTLe): A synchronization standard that allows Mark of the Unicorn's Mac-based sequencer, Performer, to lock to SMPTE through a converter which supports these standards.

Drum machine: A sample-based digital audio device that makes use of the playback capabilities of ROM (read-only) memory to reproduce carefully recorded and edited samples of individual instruments which make up the modern drum and percussion set.

Drum-pad controller: Such a controller offers the performer a larger, more expressive playing surface that can be struck either with the fingers and hands, or with the full expressiveness that can be achieved by playing with percussion mallets and drum sticks. Additionally, a drum controller will often offer extensive setup parameters.

Drum pads: The playing surface buttons which are designed into a drum machine and must be played with the fingers.

Drum-pattern editor/sequencer: A sequencer specifically designed for programming drum patterns using a straightforward, graphic format.

Frequency shift keying (FSK): A tape-to-MIDI sequencer synchronization signal which marks clock transitions by modulating two high-frequency square-wave signals onto a recorded tape track.

Hard-disk recorder: A computer-based hardware and software package specifically intended for the recording, manipulation, and reproduction of the digital audio data that resides upon hard disk and/or within the computer's own RAM.

Hardware-based sequencer: Stand-alone devices designed for the sole purpose of MIDI sequencing. These systems make use of a dedicated operating structure, microprocessing system, and memory that is integrated with top-panel controls for performing sequence-specific functions.

Interactive sequencer: A computer-based sequencer that directly interfaces with MIDI controllers and sequenced MIDI files to internally generate MIDI performance data according to a computer algorithm.

Internal sequencer: A sequencer designed into an electronic instrument that directly interfaces with the instrument's keyboard and voice structure.

Jam sync: It is the function of the jam-sync process to regenerate fresh time code onto recorded tape or to reconstruct defective sections of code

Keyboard controllers: A keyboard device expressly designed to transmit performance-related MIDI messages throughout a modular MIDI system.

Longitudinal time code (LTC): Time code which is recorded onto an audio or video cue track. LTC encodes the biphase time-code signal onto an analog audio or cue track as a modulated square-wave signal.

Memory: Used for storing important internal data, such as patch information, setup configurations, and digital waveform data.

MIDI echo: The selectable MIDI echo function is used to provide an exact copy of any information received at the MIDI in port, and route this data directly to the MIDI out/echo port.

MIDI filter: A dedicated digital device, onboard processor, or computer algorithm that allows specific MIDI messages or range of messages within a data stream to be either recognized or ignored.

MIDI implementation chart: A standardized chart that easily relates information to all of the MIDI capabilities that are supported by a specific MIDI device.

MIDI in: This port receives MIDI messages from an external source and communicates this performance, control, and timing data to the device's internal microprocessor.

MIDI interface: A device used to translate the serial message data of MIDI into a data structure that can be directly communicated both to and from a personal computer's internal operating system.

MIDI keyboard controller: A keyboard device expressly designed to transmit performance-related MIDI messages throughout a modular MIDI system.

MIDI mapper: A dedicated digital device, onboard processor, or computer algorithm that can be used to reassign the scaler value of a data byte to another assigned value.

MIDI merger: A device used to combine the data from two or more separate MIDI lines into a single MIDI data stream.

MIDI messages: These are made up of a group of related 8-bit words, which are used to convey a series of performance or control instructions to one or all MIDI devices within a system.

MIDI out: This port is used to transmit MIDI messages from a single source device to the microprocessor of another MIDI instrument or device.

MIDI patchbay: It is the function of a MIDI patchbay to selectively route MIDI data paths within a production system.

MIDI processor: Such a device may be placed within the MIDI data chain to route or alter MIDI messages. It is often capable of performing a wide range of algorithmic functions upon one or more MIDI signals.

MIDI reception modes: These modes allow a MIDI instrument to transmit or respond to channel messages in a way that can best be processed or generated by a receiving device. These modes are: Mode 1 (Omni On/Poly), Mode 2 (Omni On/Mono), Mode 3 (Omni Off/ Poly), and Mode 4 (Omni Off/ Mono).

MIDI switcher: Such a device enables the user to select between two or more MIDI controller sources without the need for the manual repatching of MIDI cables.

MIDI sync: A protocol primarily used for locking together the precise timing elements of MIDI devices within an electronic music system and operates via the transmission of real-time MIDI messages over standard MIDI cables.

MIDI thru: This port provides an exact copy of the incoming data at the MIDI in port and transmits this data out to another MIDI instrument or device that follows within the MIDI data chain.

MIDI thru box: This device is used to distribute a MIDI data source throughout a system by providing an exact copy of an incoming data signal to a number of MIDI thru ports.

MIDI time code (MTC): A system for easily and cost-effectively translating SMPTE time code into an equivalent time code that conforms to the MIDI 1.0 Specification. It also allows for time-based code and commands to be distributed throughout the MIDI chain to devices or instruments capable of understanding and executing MTC commands.

Musical Instrument Digital Interface (MIDI): A digital communications language that allows multiple electronic instruments, controllers, computers and other related devices to communicate within a connected network

Multitasking: The ability for many of the faster, more powerful personal computer's to process more than one program and/or task at a time.

Patch editor: A software-based package used to provide direct control over a compatible MIDI device, while clearly displaying each parameter setting on the monitor screen of a personal computer.

Patch librarian: A software package capable of receiving, transmitting, and often organizing patch data between one or more devices and a personal computer system.

Quantization: A timing function of a sequencer used to correct human-performance timing errors within a composition.

Sample editor: A software-based system developed to allow the personal computer to perform a variety of sample-related tasks, such as editing, digital signal processing, and system-wide sample distribution.

Sample-dump standard (SDS): A protocol for transmitting sampled digital audio and sustain loop information from one sampling device to another.

Sampler: A device capable of converting an audio signal into a digitized form, storing this digital data within its internal RAM (random-access memory) and reproducing these sounds (often polyphonically) within an audio production or musical environment.

Scoring/printing program: A software package that allows musical notation data to be entered either manually or automatically via MIDI. Once entered, the notes can be edited in an onscreen environment allowing the artist to change and configure a score using standard computer-cut and copy-and-paste techniques.

Sequencer: A sequencer is a digitally-based device used to record, edit, and output performance- and control-related MIDI messages in a sequential fashion.

Small computer systems interface (SCSI): A standardized bidirectional interface for providing a fast and direct data communications link between compatible devices at a rate of 500,000 bits/second (nearly 17 times MIDI speed) or higher.

SMPTE (Society of Motion Picture and Television Engineers) time code: A standard method of interlocking audio and video transports, which allows for the identification of an exact position on a magnetic tape by assigning a digital address to each location.

Song position pointer (SPP): SPP is used to reference a location point within a MIDI sequence (in measures) to a matching location upon an external device (such as a drum machine or tape recorder). This message provides a timing reference that increments once for every six MIDI clock messages with respect to the beginning of a composition.

Star network: A MIDI interconnection strategy allows a master controller to communicate with a number of MIDI instruments and devices (or chains of devices) over individually addressable MIDI ports. When used with a MIDI patchbay, this type of system also facilitates the direct patching of MIDI lines between devices within the network.

Standard MIDI files: A standard that has been developed to enable disk-based MIDI files to be freely interchanged between different computer-based sequencers, which use the same or different computer systems. It is made up of one or more 8-bit data streams used to encode time-stamped MIDI event data, as well as song, track, time signature, and tempo information.

Status byte: This is used as a MIDI message instruction for addressing a particular MIDI function, channel, etc.

Step time: The ability to enter notes and other events into a sequenced composition—one note at a time.

Synchronizer: A device which is used to synchronize (and often control) one or more tape or film transports within a multimedia production system.

Synchronization (sync): The occurrence of two or more events at precisely the same time. With respect to analog audio and video systems, it is achieved by interlocking the transport speeds of two or more machines.

System-exclusive (SysEx) message: SysEx data enables MIDI manufacturers, programmers, and designers to communicate customized MIDI messages between MIDI devices, as well as between these devices and MIDI computer programs. It is the purpose of these messages to give manufacturers, programmers, and designers the freedom to communicate any device-specific data of an unrestricted length, as they see fit.

System message: These messages address all devices within a system without regard to channel assignment. There are three system message types: system-common messages, system real-time messages, and SysEx messages.

Tempo mapping: A feature that allows a sequencer to be programmed to automatically change tempo within a complex song arrangement.

TTL 5 volt sync: An early form of synchronization commonly used before the adoption of MIDI. This method makes use of 5-volt clock pulses in which a swing from 0 to 5 volts represents one clock.

Universal patch editor: A software package designed to receive and transmit device-specific SysEx data to provide onscreen control over the programming functions of most (if not every) MIDI device within a system.

Vertical interval time code (VITC): A method of encoding SMPTE time code information within the video signal inside a field located outside the visible picture scan area, known as the *vertical blanking interval*.

Woodwind controllers: Instruments which are expressly designed to bring the articulation of a woodwind or brass instrument to a MIDI performance. They often provide touch sensitive keys, glide and pitch slider controls, in addition to a sensor for outputting real-time breath control over dynamics.

Bibliography

Chapter 1

Anderton, Craig, *MIDI for Musicians*, New York, NY, Amsco Publications, 1986.

Casabona, Helen and Frederick David, *Using MIDI*, Cupertino, CA, GPI Publications, 1987.

Chapter 2

De Furia, Steve, and Joe Scacciaferro, *MIDI Programmer's Handbook*, Redwood City, CA, M & T Publishing, Inc., 1989.

Boom, Michael, *Music Through MIDI*, Redmond, WA, Microsoft Press, 1987.

Casabona, Helen and Frederick David, *Using MIDI*, Cupertino, CA, GPI Publications, 1987.

Chapter 3

De Furia, Steve, and Joe Scacciaferro, *MIDI Programmer's Handbook*, Redwood City, CA, M & T Publishing, Inc., 1989.

Boom, Michael, *Music Through MIDI*, Redmond, WA, Microsoft Press, 1987.

Atari Computer Now Shipping Stacy 2 and 4—MIDI-compatible Portable Configurations for Musicians; Atari Computer's Stacy Laptop: Full MIDI Functionality in a Portable System, Atari" Computer Leads MIDI Market with Innovation, Superior System Performance, Product bulletins, Atari Corporation, January 19, 1990.

Yap, Rogelio L. Jr., "The Atari ST", *Radio Electronics*, (Feb. 1989): 99-102.

Opcode Systems Music Software and Hardware Catalog, Opcode Systems, Inc., 1990.

MIDI Time Piece and Performer 3.4 Break MIDI Channel Barrier, Press release, Mark of the Unicorn, 1990.

MSB 16/20 MIDI Switcher, Product bulletin, J.L. Cooper, 1989.

"Pelican System's MIDI Processor," *Future Music*, July 12, 1989.

MA36 MIDI Analyzer, Product bulletin, Studiomaster, Inc., Nov. 1989.

MIDI Beacon: The Essential Diagnostic Tool, Technical bulletin, Musonix, Ltd., 1990.

Alesis HR-16, Instruction manual, Alesis Corporation.

APDAlog Catalog, Fall 1989 catalog, Apple Computer, Inc.

Huber, David Miles, "The Lone Wolf of the 90s," *Studio Sound* (June, 1990).

Davies, Rick, "Musical Multitasking," *Electronic Musician*, (April, 1990): 43-52.

Chapter 4

Korg M1 Music Workstation, Press release, Korg U.S.A., Inc., 1988.

Korg M1 Music Workstation Super Guide, Product bulletin, Korg U.S.A., Inc., 1988.

Emax II 16-bit Digital Sound System, Product bulletin, E-mu Systems, Inc., 1990.

Akai Professional S1000 Series, Product bulletin, Akai Professional, 1989.

S-770 Linear Digital Sampler, Product bulletin, Roland Corporation US, 1990 US, 1990.

Proteus/proteus XR/Proteus 2 16-bit Multitimbral Digital Sound Module, Product bulletin, E-mu Systems, Inc., 1989/1990.

STUDIO 88 Plus Master Keyboard, Music Industries Corp.

HR-16 high Sample Rate/16-bit Drum Machine, Product bulletin, Alesis Corp.

DMP 18 Dynamic MIDI Pedalboard, Product bulletin, ELKA Professional.

PAD-80 Octapad II MIDI Pad Controller, Product bulletin, Roland Corporation US, 1988.

drumKAT MIDI Drum Controller, Product bulletin and specification sheet, KAT, Inc., 1989.

KAT MIDI Percussion Controller, Product bulletin and specification sheet, KAT, Inc., 1989.

GR-50 Guitar Synthesizer and GK-2 Synthesizer Driver, Product bulletin, Roland Corporation US, 1989.

Z3 Guitar Synthesizer and ZD3 Synthesizer Driver, Product bulletin, Korg Inc., 1989.

FaderMaster MIDI Command Controller, Product bulletin, J.L. Cooper Electronics, 1989.

Chapter 5

Kawai Q-80 MIDI Sequencer, Product brochure, Kawai America Corporation.

Akai ASQ10 MIDI Sequencer, Product brochure, Akai Professional, 1987.

Performer Version 3.0, 3.2 and 3.3, Product bulletin, Mark of the Unicorn, 1990.

Davies, Rick, "Mark of the Unicorn Performer 3.3 and Passport Designs Pro 4," *Electronic Musician*, (March, 1990): 102-109.

Opcode Systems Music Software and Hardware Catalog, Opcode Systems, 1990.

Accurso, Joseph, "First Takes, Voyetra Sequencer Plus Mark III Version 3.0," *Electronic Musician*, (January, 1990): 82-83.

Voyetra Catalog 90A, Voyetra Technologies, 1990.

UpBeat, Press release, Intelligent Music, November 1989.

Jam Factory, Press release, Intelligent Music, November 1989.

M Press release, Intelligent Music, November 1989.

Mpc, Voyetra Sales Brochure, Voyetra Technologies, 1990.

Sound Globs, Product brochure, Cool Shoes Software, 1990.

Q-Sheet, Users manual, Digidesign, Inc., 1988.

Chapter 6

Anderton, Craig, *MIDI for Musicians*, New York, NY, Amsco Publications, 1986.

Casabona, Helen and Frederick David, *Using MIDI*, Cupertino, CA, GPI Publications, 1987.

GenEdit Universal MIDI Editor, Hybrid Arts, Inc.

Product Catalog, Voyetra Technologies, 1990.

Music and Software Catalog, Opcode Systems, Inc., 1990.

Galaxy Universal Librarian, Users manual, Opcode Systems, Inc., 1990.

Chapter 7

MusicPrinter Plus Version 3.0, Product bulletin, Temporal Acuity Products, Inc., 1989.

MusicProse, Product bulletin, Coda Music Software, 1989.

Personal Composer 3.3, Owners manual and product bulletin, Personal Composer, 1990.

Performer Version 3.4, Addendum, Mark of the Unicorn, 1990.

Chapter 8

Rane MPE 28 MIDI Programmable Equalizer, Product bulletin, Rane Corporation, 1989.

Yamaha DEQ7, News release, Yamaha International Corporation, 1986.

PRO HUSH, Product bulletin, Rocktron Corporation, 1990.

MIDIVERB III, Product bulletin, Alesis Studio Electronics, 1989.

ART SGE MACH II, Product bulletin, Applied Research & Technology, Inc., 1989.

Multi-Effects, MIDI and the Lexicon LXP-1, Applications Bulletin, 1989, and Lexicon Effects System, Product bulletin, AKG Acoustics, Inc., 1989.

Lexicon LXP-5 Digital Multi-Effects Processor, Product bulletin, AKG Acoustics, Inc., 1989.

Lexicon Pcm 70 Digital Effects Processor, Product bulletin, AKG Acoustics, Inc., 1989.

GP-16 Digital Guitar Effects Processor, Product bulletin, Roland Corporation US, 1989.

Chapter 9

Huber, David Miles, *Audio Production Techniques for Video*, Indianapolis, Howard W. Sams & Company, 1987.

Huber, David Miles and Runstein, Robert, *Modern Recording Techniques, Third Edition*, Indianapolis, Howard W. Sams & Company, 1989.

MIDI Time Code and Cueing - Detailed Specification, (Supplement to MIDI 1.0), 12 February 1987.

Lehrman, Paul D., "The Future of MIDI Time Code," *Recording Engineer/Producer*, (October, 1987): 106-113.

PPS-1 Version Three, PPS-100, Product bulletin, J.L. Cooper, 1989.

Opcode Releases the Timecode Machine - SMPTE to MIDI Time Code Converter, Press release, Opcode Systems, December, 1987.

Casabona, Helen and Frederick David, *Using MIDI*, Cupertino, CA, GPI Publications, 1987.

Performer 3.0, Software manual, Mark of the Unicorn, 1989.

Anderton, Craig, "The MIDI Connection: Automated Sessions for Acoustic Musicians," *Electronic Musician*, (March, 1990): 43-48, 62-65.

Chapter 10

Huber, David Miles, *Random Access Audio*, Menlo Park, CA, Digidesign, Inc., 1990.

Sound Designer II, Software manual, Digidesign, Inc., 1989.

Huber, David Miles and Runstein, Robert, *Modern Recording Techniques, Third Edition*, Indianapolis, Howard W. Sams & Company, 1989.

Chapter 11

Sequencer Plus Mark II, Software manual, Voyetra Technologies, 1989.

Q-Sheet User's Manual, Digidesign, Inc., 1988.

Lehrman, Paul D., "Control of Effects, One Designer's Viewpoint," *Recording Engineer/Producer*, (November, 1988).

Performer Version 3.2 Highlights, Product Bulletin, Mark of the Unicorn, 1989.

Mixmate Automation Mixdown System, Product bulletin, J.L. Cooper Electronics, 1989.

MAGI-II Console Automation System, Product bulletin, J.L. Cooper Electronics, 1989.

Akai MPX820 Mixer, Product bulletin, Akai professional, 1989.

Huber, David Miles and Runstein, Robert, *Modern Recording Techniques, Third Edition*, Indianapolis, Howard W. Sams & Company, 1989.

Index

D

E

F

G

H

I

J

K

L

M

N

O

T

U

V

W

X-Z